Power Systems

For further volumes:
http://www.springer.com/series/4622

Mathias Uslar, Michael Specht, Christian Dänekas,
Jörn Trefke, Sebastian Rohjans, José M. González,
Christine Rosinger, and Robert Bleiker

Standardization in Smart Grids

Introduction to IT-Related Methodologies, Architectures and Standards

Springer

Authors
Dr.-Ing. Mathias Uslar
OFFIS – Institut für Informatik
Oldenburg
Germany

Dipl.-Inform. (FH) Michael Specht
OFFIS – Institut für Informatik
Oldenburg
Germany

Dipl.-Inform. Christian Dänekas
OFFIS – Institut für Informatik
Oldenburg
Germany

Dipl.-Inform. Jörn Trefke
OFFIS – Institut für Informatik
Oldenburg
Germany

Dipl.-Inform. Sebastian Rohjans
OFFIS – Institut für Informatik
Oldenburg
Germany

Dipl.-Wirt.Inform. José M. González
OFFIS – Institut für Informatik
Oldenburg
Germany

Dipl.-Inform. Christine Rosinger
OFFIS – Institut für Informatik
Oldenburg
Germany

Dipl.-Inform. Robert Bleiker
OFFIS – Institut für Informatik
Oldenburg
Germany

ISSN 1612-1287 e-ISSN 1860-4676
ISBN 978-3-642-34915-7 e-ISBN 978-3-642-34916-4
DOI 10.1007/978-3-642-34916-4
Springer Heidelberg New York Dordrecht London

Library of Congress Control Number: 2012952596

Printed on acid-free paper

Springer is part of Springer Science+Business Media (www.springer.com)

"Any sufficiently advanced technology is indistinguishable from magic."

Arthur C. Clarke

Foreword

Coping with the future of electric energy supply for Europe is one of the key goals of the 20-20-20 targets defined by the European Commission. To reach these targets, it is mandatory to transform the existing power infrastructure into a smart, decentralized, resource-efficient, emission-efficient, yet still dependable and affordable system-of-systems. This is a large challenge that the utilities, vendors, regulators and, of course, customers have to deal with. In this context, the development of the Smart Grid has become a central point of attention. The European Commission has viewed the evolution towards the smart grid as a very complex and multi-facetted transformation process and, at an early stage, has established a structure to address this. The European Technology Platform for Electricity Networks of the Future, also called SmartGrids ETP, is the key European forum for the crystallization of policy, technology and research and development pathways for the smart grids sector, as well as the link with other EU related initiatives. In a complex smart grid, the value chain ranges from generation to appliances. Besides the regulatory and market aspects, the technical level has to deal with knowledge from different sectors, multiple disciplines and issues of technical system integration and interoperability. These questions are typically addressed and resolved by the definition and usage of (technical) standards for processes, data models, functions and communication links. Standardization is a key issue for smart grids and the standards landscape is obviously very large and complex. This is why the three European Standards Organisations ETSI, CEN and CENELEC have first created a Joint Working Group (JWG which was the first harmonized effort in Europe to bring together the needed disciplines and experts). The JWG produced in May 2011, a report that outlines Europe's standardization views in the area of smart grids, taking due account of existing global activities. Based in particular on the JWG results, the European Commission has issued the M/490 Standardization Mandate towards the European Standardisation Organisations (ESOs) to support the deployment of the European Smart Grid. The focal point for the ESO's response to M/490 is the CEN, CENELEC and ETSI Smart Grids Co-ordination Group (SG-CG). The main role of the SG-CG is to define a modern approach to standardization that will guarantee that the EU Smart Grid standardization will be undertaken in a coherent manner over the next years in the appropriate technical committees. The approach taken is centered around - first - the development of a consistent methodology for the development of use cases, reference architectures, communication technologies, data models and information security models; and - second - the selection of the appropriate existing standards and the identification of new ones whose development by the standardization community is required. This book provides an overview on the various building blocks and standards identified as the most prominent ones in the JWG report as well as by the SG-CG groups. It also introduces the Smart Grid Architecture Model (SGAM) for utilities, as well as future standards for market communications, electric vehicles and future industrial automation. As the convener of the SG-CG Reference Architecture Working Group, I welcome the initiative to come up with a textbook providing meaningful introductions into the various standards for smart grid arising from the recommendations of the M/490 mandate. I am confident that this textbook will be the best possible introduction to readers wanting a sound and clear access to a very interesting, yet very complex, topic.

Sophia-Antipolis, September 2012

Emmanuel Darmois,
Convener of the M/490 Reference Architecture Work Group,
Alcatel-Lucent

Foreword

During the first International Conference on the Integration of Renewable Energy Sources and Distributed Energy Resources held in December 2004, industrial stakeholders and the research community suggested the creation of an European Technology Platform for the Electricity Networks of the Future. In April 2006, the Advisory Council of this European Technology Platform presented its so called Vision document for Smart Grids. The Vision, for both transmission and distribution networks, is driven by the combined effects of market liberalization, the change in generation technologies to meet environmental targets and the future uses of electricity. Together with the Strategic Research Agenda, published in 2007, it described the main areas to be investigated, technical and non-technical, in the short-medium term in Europe. Since then, these documents have inspired several Research and Development programs within the EU and National institutions. One particular aspect which has been identified by the various demo projects funded by the commission is, besides the highly important market and regulation aspects, the aspect of technical interoperability and standardization. In March 2011, this issue was addressed by giving the M/490 mandate to the relevant European Standardization Organizations (ESOs).

The objective of this mandate is to develop or update a set of consistent standards within a common European framework. As well as integrating a variety of digital computing and communication technologies and electrical architectures, associated processes, and services, that will achieve interoperability and will enable or facilitate the implementation in Europe of the different high level Smart Grid services and functionalities as defined by the Smart Grid Task Force, which should be flexible enough to accommodate future developments. Building, Industry, Appliances, and Home Automation are out of the scope of this mandate; however, their interfaces with the Smart Grid and related services have to be treated under this mandate.

Within this book, the authors refer to those aforementioned goals from the excellent Smart Grid Coordination Group reports. Within the mandate, use cases for the European Smart Grid were collected and harmonized by the Sustainable Processes (SP) Group according to the IEC PAS 62559 methodology. The first set of standards (FSS) group created, based on previous joint working group (JWG) work, a meaningful list of core standards for the Smart Grid. The two most prominent ones, CIM and IEC 61850, among others, are introduced in this book. Additionally, the SGIS's work on security for Smart Grid is reflected in the chapter dealing with the most important Smart Grid security aspects. The Smart Grid Architecture Model (SGAM), one of the core aspects form the Reference Architecture Working Group and its origins and applications is described in three chapters of this work. In addition, links to other mandates like M/468 on Electric Vehicles and future applications like certification, testing for standards, possible market communication profiles and automation standards like OPC UA are discussed.

The authors provide an introductory text book on the various aspects of how to deal with the future and existing Smart Grid communication standards. I hope a lot of readers will benefit from this material and de-mystify certain technical aspects on how to achieve the goal of a sustainable and cost-efficient Smart Grid for Europe. All the best to this first edition.

Essen, September 2012

Thomas Theisen,
Head of New Technology
RWE Deutschland AG

Preface

One of the predominant topics in the domain of the emerging Smart Grid can be seen in standardization. With the combination of existing protection and automation technology with upcoming ICT-based solutions, different interoperability issues arise when technologies have to be combined in the infrastructure. Standards have proven to be one of the most striking solutions to actually cope with this topic. Since 2008, this topic has gained much attention in various political and technical agendas.

At OFFIS - Institute for Information Technology, we started working on the very topic in 2004. In 2009, the group "Interoperability and Standardization" was founded as one part of the R&D Division "Energy". The focus of the work is on the meaningful application of software engineering and interoperability research for the utility domain, mainly utilizing and extending smart grid standards. This book provides an overview on our portfolio of the various research trends and standards applied. The individual chapters provide short overviews and emerging Smart Grid standards as well as derived methodologies, which can be applied to Smart Grid development.

Focal topics are the application of the IEC 62559 IntelliGrid methodology for use case management, the use of the SGAM for the EU mandate M/490 and the most important IEC standards CIM and IEC 61850. In addition, future trends and emerging standards are introduced. The editors and authors hope that this book will prove useful as both ana introductory textbook for people trying to get first hand and condensed knowledge on Smart Grid standardization with a focus on ICT as well as to have a reference textbook dealing with the various standards to be applied in Smart Grids.

Oldenburg, September 2012 The Authors

Acknowledgements

The authors would like to thank all the partners who have contributed to this book through feedback on the trainings given by trainers from OFFIS. Without this, we would not have had the opportunity to write this textbook coming up with the essentials on Smart Grid standardization from a communication perspective. In addition, the work carried out in the mandate M/490 to CEN/CENELEC/ETSI makes for a meaningful step forward in Smart Grid standardization. This book hopes to reflect the ideas of the preliminary as well as the final reports and wants to thank the numerous, most of the time unnamed authors and experts of the work packages in the reports.

Contents

Part I Basics and Introduction

1 Introduction and Smart Grid Basics 3
Mathias Uslar
1.1 Smart Grid—What Is It? 3
1.2 General Motivation for Standardization in Smart Grids 6
1.3 Internationally Recommended Core Standards for Communications and Data Modeling 6
1.4 The EU Mandate M/490 to CEN/CENELEC and ETSI 8
1.5 Conclusion 11
References 12

Part II Requirements and Architectures

2 Requirements Engineering for Smart Grids 15
Christian Dänekas, José M. González
2.1 Motivation 15
2.2 Requirements Engineering Concepts and Process Integration 16
2.3 Managing Smart Grid Requirements Regarding Interoperability Aspects 19
2.4 Architectural Viewpoints towards Smart Grid Requirements 21
2.5 Exemplary Requirements Analysis for Advanced Metering Infrastructure Using the Smart Grid Architectural Model 23
2.6 Management of Long-Term Smart Grid Requirements 31
2.7 Summary and Outlook 35
References 36

3 IEC/PAS 62559-Based Use Case Management for Smart Grids 39
Jörn Trefke, José M. González, Christian Dänekas
3.1 Introduction to Use Case Modeling for Smart Grids 39
3.2 Requirements for Smart Grid Use Case Descriptions 40

3.3 Smart Grid Use Case Methodology 42
3.4 Summary and Outlook on Future Work 56
References 57

4 Development of Smart Grid Architectures 59
Jörn Trefke, Christian Dänekas
4.1 Motivation for Architecture Development 59
4.2 On Architecture 60
4.3 Viewpoints for Enterprise Architecture 65
4.4 An Approach for Enterprise Architecture Development and Its Management 67
4.5 Conclusion and Outlook 75
References 77

5 Management of Information Models in the Energy Sector 79
José M. González, Jörn Trefke
5.1 Smart Grids and Challenges for Enterprises 79
5.2 Introduction to Information Models 81
5.3 Information Sources for Requirements Analysis within the German Energy Sector 85
5.4 Information Systems in the Energy Sector 87
5.5 The Energy Reference Model Catalog 90
5.6 Summary and Outlook 94
References 94

Part III Standards and Applications

6 ICT and Energy Supply: IEC 61970/61968 Common Information Model 99
Michael Specht, Sebastian Rohjans
6.1 Introduction and History 99
6.2 Data Models 101
6.3 Profiles 103
6.4 Serializations 104
6.5 Component Interface Specifications (CIS) 110
6.6 Interface Reference Model (IRM) 111
6.7 Tooling 111
6.8 Conclusion and Outlook 113
References 114

7 Automation for the Smart Grid: IEC 61850 - Substation Automation and DER Communication 115
Mathias Uslar, Robert Bleiker
7.1 Introduction to the IEC 61850 Standard Family 115
7.2 History and Overview 116
7.3 The Architecture 117

7.4 Parts of the Standard Family ... 120
7.5 Conclusion and Outlook ... 127
References ... 127

8 Smart Grid Security: IEC 62351 and Other Relevant Standards ... 129
Christine Rosinger, Mathias Uslar
8.1 Introduction and Motivation ... 129
8.2 Previous Incidents and Attack Patterns ... 130
8.3 Recommended Security Standards ... 131
8.4 Security Metrics ... 142
8.5 Security Patterns ... 143
8.6 Conclusion ... 144
References ... 145

9 Testing in the Smart Grid: Compliance, Conformance and Interoperability ... 147
Robert Bleiker, Michael Specht
9.1 Principles of Testing ... 147
9.2 IEC 61850 Testing ... 152
9.3 Common Information Model (CIM) Testing ... 156
References ... 160

10 Standards in the Electro Mobility Domain—Vehicle 2 Grid ... 163
Michael Specht, Christine Rosinger
10.1 Introduction ... 163
10.2 Evolutionary Steps ... 163
10.3 Scenarios ... 164
10.4 Security for the Electro Mobility Domain ... 172
10.5 Existing Standards Relevant for ICT in the Electro Mobility Domain ... 174
10.6 Conclusion and Outlook ... 176
References ... 176

11 Smart Metering in the European Context ... 179
Michael Specht
11.1 Introduction ... 179
11.2 CEN/CENELEC/ETSI Smart Meters Coordination Group Report for EU-Mandate M/441 ... 180
11.3 IEC 62056 DLMS/COSEM ... 182
11.4 Harmonization of DLMS and CIM ... 183
11.5 Smart Message Language ... 184
11.6 Metering Bus ... 184
11.7 ANSI C12 ... 185
11.8 KNX ... 185
11.9 ZigBee Smart Energy Profile ... 186

11.10 Conclusion and Outlook 187
References 188

Part IV Future Applications and Outlook

12 OPC UA: An Automation Standard for Future Smart Grids 191
Sebastian Rohjans, Michael Specht
12.1 Introduction and History 191
12.2 Information Modeling 194
12.3 Communication Services 196
12.4 Technology Mappings 198
12.5 Profiles 199
12.6 Security 199
12.7 Power Domain-Specific Data Modeling 201
12.8 Conclusion and Outlook 207
References 208

13 Market Communication 211
José M. González, Michael Specht
13.1 Market Communication and the Need for IT Standards 211
13.2 IT Standards and Standards Developing Organizations for Market Communication 213
13.3 Summary 225
References 227

14 Looking Ahead: The Future of Smart Grid Communications and Standardization 229
Mathias Uslar
14.1 The Good 229
14.2 The Bad 230
14.3 The Ugly 231
14.4 Recommendations and Trends 233
References 233

A CIM Package Description 235

B CIM RDF Topology 237

C Exemplary Use Case According to an Extended IEC/PAS 62559 Template 239

Index 249

List of Contributors

Robert Bleiker
OFFIS – Institute for Information Technology, Escherweg 2,
26121 Oldenburg, Germany
e-mail: bleiker@offis.de

Christian Dänekas
OFFIS – Institute for Information Technology, Escherweg 2,
26121 Oldenburg, Germany
e-mail: daenekas@offis.de

Jose M. Gonzalez
OFFIS – Institute for Information Technology, Escherweg 2,
26121 Oldenburg, Germany
e-mail: gonzalez@offis.de

Sebastian Rohjans
OFFIS – Institute for Information Technology, Escherweg 2,
26121 Oldenburg, Germany
e-mail: rohjans@offis.de

Christine Rosinger
OFFIS – Institute for Information Technology, Escherweg 2,
26121 Oldenburg, Germany
e-mail: christine.rosinger@offis.de

Michael Specht
OFFIS – Institute for Information Technology, Escherweg 2,
26121 Oldenburg, Germany
e-mail: specht@offis.de

Jörn Trefke
OFFIS – Institute for Information Technology, Escherweg 2,
26121 Oldenburg, Germany
e-mail: trefke@offis.de

Mathias Uslar
OFFIS – Institute for Information Technology, Escherweg 2,
26121 Oldenburg, Germany
e-mail: uslar@offis.de

Acronyms

3GPP	3rd Generation Partnership Project
AC	Alternating Current
ACSI	Abstract Communication System Interface
AD	Architecture Description
ADI	Analyzer Device Integration
ADM	Architecture Development Method
ADR	Automated Demand Response
AE	Alarms and Events
AES	Advanced Encryption Standard
AMI	Advanced or Automated Metering Infrastructure
AMR	Automated Meter Reading
ANSI	American National Standards Institute
APDU	Application Protocol Data Units
API	Application Programming Interface
ARIS	Architecture of Integrated Information Systems
ASHRAE	American Society of Heating, Refrigerating and Air-Conditioning Engineers
AUTOSAR	AUTomotive Open System ARchitecture
BACnet	Building Automation and Control Networks
BDEW	Bundesverband der Energie- und Wasserwirtschaft e. V. (engl.: German Association of Energy and Water Industries)
BITKOM	Bundesverband Informationswirtschaft, Telekommunikation und neue Medien (engl.: German Association for Information Technology, Telecommunication and New Media)
BK	Beschlusskammer (engl.: Ruling Chamber)
BMWi	Bundesministeriums für Wirtschaft und Technologie (engl.: German Federal Ministry of Economics and Technology)
BNetzA	Bundesnetzagentur (engl.: German Federal Network Agency for Electricity, Gas, Telecommunications, Post and Railway)
BOV	Business Operations View
BPMN	Business Process Model and Notation

BSI	Bundesamt für Sicherheit in der Informationstechnik (engl.: German Federal Ministry of IT Security)
CAN	Controller Area Network
CC	Common Criteria or Creative Commons (License)
CCAPI	Control Center Application Programming Interface
CCTS	Core Components Technical Specification
CDA	Common Data Attribute
CDC	Common Data Class
CDPSM	Common Distribution Power System Model
CEDEC	Confédéation Européenne des Entreprises Locales d'Energie (engl.: European Federation of Local Energy Companies)
CEN	European Committee for Standardization
CENELEC	European Committee for Electrotechnical Standardization
CHP	Combined Heat and Power
CI	Communication Interface
CIM	Common Information Model
CIMbaT	CIM-based Transformation (software tool for mapping CIM to OPC-UA)
CIMug	CIM users group
CIP	Critical Infrastructure Protection
CIS	Common Interface Specifications
CLC	*C*ENE*L*E*C*
CLS	Controllable Local System
CME	CIM Market Extension
COM	Component Object Model
COSEM	COmpanion Specification for Energy Metering
CPSM	Common Power System Model
CRM	Customer-Relationship-Management
CSMS	Cyber Security Management Systems
CuS	Customer Switching
CUST	Customer
CWM	Chronos Web Modeller
DA	Data Access
DCIM	CIM for Distribution
DCOM	Distributed COM
DER	Distributed Energy Resources
DG	Distributed Generation
DI	Device Integration
DigSig	Guidelines for Digital Signatures
DIN	Deutsches Institut für Normung e. V. (engl.: German Institute for Standardization)
DKE	DKE Deutsche Kommission Elektrotechnik Elektronik Informationstechnik im DIN und VDE (engl.: German Commission for Electrical, Electronic & Information Technologies)
DLMS	Device Language Message specification

DMS	Distribution Management System
DNP	Distributed Network Protocol
DOM	Document Object Model
DoS	Denial-of-Services
DR	Demand Response
DSM	Demand Side Management
DSO	Distribution System Operator
DTR	Draft Technical Report
DVGW	Deutscher Verein des Gas- und Wasserfaches
EA	Enterprise Architect
EAI	Enterprise Application Integration
EAM	Enterprise Architecture Management
EASEE-gas	European Association for the Streamlining of Energy Exchange-gas
ebIX	energy Business Information eXchange
ebXML	Electronic Business using XML
EC	European Commission
ECAN	ENTSO-E Capacity Allocation and Nomination
ECC	ENTSO-E Core Components
ECE	Economic Commission for Europe
eCM	Electronic Confirmation and Matching
EDGE	Enhanced Data Rates for GSM Evolution
EDI	Electronic data interchange
EDIFACT	United Nations Electronic Data Interchange For Administration, Commerce and Transport
EDIXML	EDI in XML
EDL	Exchange Data Language
EDM	Energy Data Management
EDSO	European Distribution System Operators
EEGI	European Electricity Grid Initiative
EFET	European Federation of Energy Traders
EG	Expert Group
EHS	European Home Systems (Protocol)
EIB	European Installation Bus
EIC	ENTSO-E Energy Identification Coding Scheme
EMB	Enterprise Message Bus
EMD	Exchange of Metered Data
EMIX	Energy Market Information Exchange
EMM	ENTSO-E Modeling Methodology
EMS	Energy Management System
EN	European Norm
energy RMC	energy Reference Model Catalog
ENTSO-E	European Network of Transmission System Operators for Electricity
EnWG	Energy Industry Act

EPC	Event-driven Process Chain
ePM	Electronic Position Matching
EPRI	Electric Power Research Institute
ERCOT	Electric Reliability Council of Texas
ERM	Entity-Relationship-Model
ERRP	ENTSO-E Reserve Resource Process
ESB	Enterprise Service Bus
ESCoRTS	*E*uropean Network for the *S*ecurity of *Co*ntrol and *R*eal *T*ime *S*ystems
eSM	Electronic Settlement Matching
ESO	European Standardization Organization
ESP	ENTSO-E Settlement Process
ESS	ENTSO-E Scheduling System
ETC	ebIX Technical Committee
ETP	European Technology Platform
ETSI	European Telecommunications Standards Institute
ETSO	European Transmission System Operators
EU	European Union
EURELECTRIC	The Union of the Electricity Industry-Eurelectric
EV	Electrical Vehicle
EVU	Energieversorgungsunternehmen (engl.: Utility)
FACTS	Flexible Alternating Current Transmission System
FC	Functional Constraint
FEG	Future Energy Grid
FERC	Federal Energy Regulatory Commission
FIBEX	Field Bus Exchange Format
FRM	Functional Reference Model
FSS	First Set of Standards (working group within SG-CG
FTP	File Transfer Protocol
GA	EFET Standard Documentation General Agreements
GABi Gas	Grundmodell der Ausgleichsleistungen und Bilanzierungsregeln im deutschen Gasmarkt (engl.: Basic Model of the German Federal Network Agency for Compensatory Energy and Balancing Rules in the Gas Sector)
GasNZV	network access ordinances for electricity
GAWANIS	GAs- and WAter Network Information Systems
GDA	Generic Data Access
GeLi Gas	Supplier Switching Processes for gas
GES	Generic Eventing and Subscription
GID	Generic Interface Definition
GIS	Geographic Information System
GO	Grid Operations
GOOSE	Generic Object Oriented Substation Events
GPKE	Supplier Switching Processes for electricity
GPRS	General packet radio service

GSE	Generic Substation Events
GSM	Global System for Mobile Communications
GWAC	GridWise Architectural Council
HAN	Home Area Network
HBES	Home and Building Electronic System
HDA	Historical Data Access
HMAC	Hash Message Authentication Code
HMI	Human Machine Interface
HSDA	High-Speed Data Access
HTML	HyperText Markup Language
HTTP	HyperText Transfer Protocol
HTTPS	Secure HTTP
IACS	Industrial Automation and Control System
IBM	International Business Machines Corporation
ICD	IED Capability Description
ICT	Information and Communication Technologies
ID	Identifier
IDS	Intrusion Detection System or Integrierte Datenverarbeitungs-Systeme (engl.: Integrated Data Processing Systems)
IEC	International Electrotechnical Commission
IED	Intelligent Electronic Device
IEEE	Institute of Electrical and Electronics Engineers
INVOIC	Billing messages between network operator and supplier
IOP	Interoperability
IP	Internet Protocol
IRM	Interface Reference Model
IS	Information System or International Standard
ISA	International Society of Automation
ISMS	Information Security Management Systems
ISO/OSI	International Organization for Standardization/Open Systems Interconnection
ISO	International Organization for Standardization
IT	Information Technologies
ITIL	IT Infrastructure Library
ITU	International Telecommunication Union
JISC	Japanese Industrial Standards Committee
JMS	Java Messaging System
JRE	Java Runtime Environment
JWG	Joint Working Group
KPI	Key Performance Indicator
KPMG	Professional services company
KTH	Kungliga Tekniska högskolan (engl.: (Swedish) Royal Institute of Technology)
LAN	Local Area Network
LD	Logical Device

LIN	Local Interconnection Network
LMN	Local Metrological Network
LN	Logical Node
LPHD	Logical Node Physical Device
LTE	Long Term Evolution
LV	Low Voltage
MaBIS	Marktregeln für die Durchführung der Bilanzkreisabrechnung Strom (engl.: Market rules for conducting settlement area billing in the electricity sector)
MAC	Message Authentication Code
MADES	ENTSO-E Market Data Exchange Standard
MDM	Meter Data Management
MMS	Manufacturing Messaging Specification
MOST	Media Oriented Systems Transport
MoU	Memorandum of Understanding
MRASCo	MRA Service Company
mRID	Message Resource Identifier
MTTR	Mean Time to Recovery
MUC	Multi-Utility Controller
MV	Medium Voltage
NDR	Naming and Design Rules
NEMA	(US) National Electrical Manufacturers Association
NERC	North American Electric Reliability Council
NIST	National Institute of Standards and Technology
NISTIR	NIST Interagency Report
NSM	Network and System Management
NWIP	New Work Item Proposal
OASIS	Organization for the Advancement of Structured Information Standards
OEM	Original Equipment Manufacturer
OGEMA	Open Gateway Energy Management Alliance
OLE	Object Linking and Embedding
OMS	Outage Management System
OO	Object-oriented
OPC	OLE for Process Control
OPC-UA	OPC - Unified Architecture
OS	Organization and Structure
OSI	Open Systems Interconnection
OTC	Over The Counter
OWL	Web Ontology Language
PAP	Priority Action Plan
PAS	Publicly Available Specifications
PC	Project Committee
PEV	Plug-in Electric Vehicles
PHD	Physical Device

PIM	Platform Independent Model
PKI	Public-Key-Infrastructure
PLC	Power Line Communications or Programmable Logic Controller
PMU	Phasor Measurement Unit
PP	Protection Profile
PPP	Point-to-Point Protocol
PTB	Physikalisch-Technische Bundesanstalt (engl.: German Physical -Technical Federal Institute)
QoS	Quality of Service
RA	Reference Architecture
RBAC	Role-Based Access Control
RDF	Resource Description Framework
RMC	Reference Model Catalog
RMR	Remote Meter Reading
ROI	Return of Investment
RUP	Rational Unified Process
SA	Substation Automation
SAP	Systemanalyse und Programmentwicklung (engl.: System Analysis and Program Development) – a software company
SC	Subcommittee
SCADA	Supervisory Control and Data Acquisition
SCC	Standards Coordinating Committee
SCL	System Configuration Language, former Substation Configuration Language
SCSM	Specific Communication Service Mapping
SDK	Software Development Kit
SDO	Standard Developing Organization
SE	Societal and Environmental
SEC	Security
SEI	Software Engineering Institute
SG	Smart Grid or Strategic Group
SGAC	Smart Grid Advisory Committee
SGAM	Smart Grid Architectural Model
SG-CG	EU Smart Grid Coordination Group
SG-CG	Smart Grid Coordination Group
SGIS	Smart Grid Information Security (working group of SG-CG)
SGMM	Smart Grid Maturity Model
SIA	Seamless Integration Reference Architecture
SIDM	System Interfaces for Distribution Management
SMB	Standardization Management Board
SM-CG	Smart Meters Coordination Group
SML	Smart Message Language
SMR	Strategy, Management, and Regulatory
SMV	Sampled Measured Values
SOA	Service-oriented architecture

SOAP	SOAP Simple Object Access Protocol
SP	Special Publication or Sustainable Processes (working group of SG-CG)
SPEC	Specification of DIN, e.g., PAS, CEN Workshop Agreement, Pre-Standard or technical report
SS	Substation
SSL	Secure Sockets Layer
StromNZV	network access ordinances for electricity
TASE	Telecontrol Application Service Element
TC	Technical Committee
TCO	Total Cost of Ownership
TCP/IP	Transmission Control Protocol and Internet Protocol
TDL	Table Description Language
TECH	Technology
TF	Task Force
TLS	Transport Layer Security
TOE	Target of Evaluation
TOGAF	The Open Group Architecture Framework
ToR	Terms of Reference
TR	Technical Report
TS	Technical Specification
TSDA	Time Series Data Access
TSO	Transmission System Operator
UA-SC	UA Secure Conversation
UC	Use Case
UCMR	Use-Case-Management-Repository
UDP	User Datagram Protocol
UML	Unified Modeling Language
UMM	UN/CEFACT Modeling Methodology
UMTS	Universal Mobile Telecommunications System
UN/CCTS	UN/CEFACT Core Components Technical Specification
UN/CEFACT	UN Centre for Trade Facilitation and E-business
UPS	Uninterruptible Power Systems
URI	Uniform Resource Identifier
URL	Uniform Resource Locator
UTILMD	Message regarding data exchange during supplier switching
V2G	Vehicle-to-grid
VCI	Value Chain Integration
VDE	Verband der Elektrotechnik Elektronik Informationstechnik e. V. (engl.: German Association for Electrical, Electronic and Information Technologies)
VPP	Virtual Power Plant
WAM	Work and Asset Management
WAN	Wide Area Network
WG	Working Group

WiFi	Technology to wirelessly exchange data over a computer network. WiFi and WLAN are often used synonymously.
WiM	Wechselprozesse im Messwesen (engl.: change processes in metrology)
WLAN	Wireless LAN
WMS	Workforce Management System
WPP	Wind Power Plant
WS	Web Service
XML	Extensible Markup Language
XSD	XML Schema Definition

Part I
Basics and Introduction

Chapter 1
Introduction and Smart Grid Basics

Mathias Uslar

Abstract. This chapter is going to provide a short overview on the basic definitions of smart grid as well as an introduction to recommendations from the most relevant international initiatives on standardizing this very topic. Its focus is on providing an overview on the latest European Smart Grid initiatives on Communication Standards from the Smart Grid Coordination Group SG-CG for the mandate M/490.

1.1 Smart Grid—What Is It?

The Smart Grid is one of the dominating topics discussed today in the energy domain. Due to previous experiences as well as several national and international studies and roadmaps like [3], [10] or [7], it is generally accepted, that an appropriate Information and Communication Technologies (ICT) infrastructure is needed to control the future power type of power transmission and distribution grid and gather relevant data. Furthermore, the use of standards within this future infrastructure is indispensable as outlined by the aforementioned roadmaps and studies in order to reach a proper interoperability level.

A lot of new functions, services and use cases arise and stakeholders within the (electric) Smart Grid have to cope with those new challenges, which are, amongst others, highly focusing on interoperability of the components to be integrated [10].

The term Smart Grid is one of the most frequently used words in the energy domain in the last few years. To establish a common basis, we provide the definition of Smart Grid for the context of this book in this section.

The term "Smart Grid" (an intelligent energy supply system) comprises the networking and control of intelligent generators, storage facilities, loads and network operating equipment in power transmission and distribution networks with the aid

Mathias Uslar
OFFIS – Institute for Information Technology, Escherweg 2, 26121 Oldenburg, Germany
e-mail: `uslar@offis.de`

M. Uslar et al.: *Standardization in Smart Grids*, POWSYS, pp. 3–12.
DOI: 10.1007/978-3-642-34916-4_1

of Information and Communication Technologies. The objective is to ensure sustainable and environmentally sound power supply by means of transparent, energy - and cost-efficient, safe and reliable system operation.

The above definition is used in the German "Deutsche Kommission Elektrotechnik Elektronik Informationstechnik im DIN und VDE (DKE)" SMART.GRID mirror committee according to the International Electrotechnical Commission (IEC) Standardization Management Board Strategic Group 3 (SMB SG 3) "Smart Grid" and will be the very core of our definition like in [4]. Different power generation structures in various countries lead to the fact that a Smart Grid cannot be defined as a only one type and amount of attributes being possible. Regulation, different utility requirements, and, of course, natural resources are resulting in different requirements for the Smart Grid and the electric distribution and transmission grid. But there are not only widely differing requirements in the generation part of the grid, in addition, also consumer structures have various characteristics in terms of energy prices or urban population density. Thus, the coordinated balance between power generation and consumption may be different in relation to the corresponding location. There are central aspects which can be defined as follows:

- It is necessary to integrate more variable generation of distributed generation and storage options.
- A holistic, intelligent energy supply system called Smart Grid comprises active power distribution and transmission with new ICT-based technologies
- Distributed and decentralized grid with distributed coordination intelligence and data aggregation management is one of the core concepts
- Customer involvement in form of a so called prosumer (consumer and possible producer) residing in a smart building with electric vehicles and the necessary ICT-based equipment to support the decentralized decision making, bidirectionally communication with the utility, and, most of the time, smart-metering

Also, there are further international definitions which could be consulted to both define and coin the term Smart Grid like the ones following in the next paragraphs.

The European Technology Platform (ETP) "Smart Grids" [8] has defined a strategic vision for a so called "transition" to the Smart Grid. The use of common technical standards and protocols to achieve open access, interoperability and vendor independence of the components is one of the core demands. Also, bidirectional connection for both communication and electricity flow is considered as a key feature. From the viewpoint of the ETP "Smart Grids", the Smart Grid is defined as follows (see figure 1.1):

A Smart Grid is an electricity network that can intelligently integrate the actions of all users connected to it—generators, consumers—and those that do both—in order to efficiently deliver sustainable, economic and secure electricity supplies.

The American National Institute of Standards and Technology (NIST)[1] has another definition as follows [5]:

[1] http://www.nist.gov/smartgrid/index.cfm

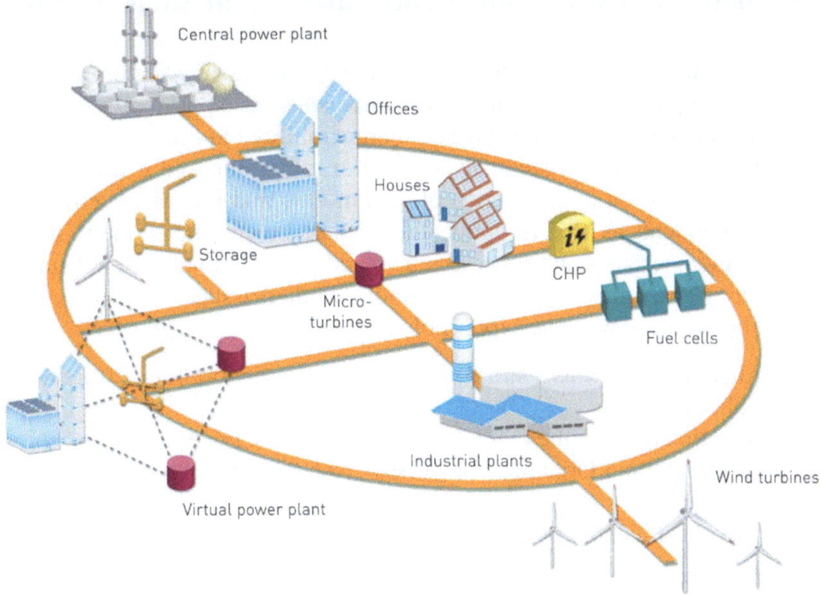

Fig. 1.1 The European Union's Smart Grid vision of the European Technology Platform SG [8]

The term "Smart Grid" refers to a modernization of the electricity delivery system so it monitors, protects and automatically optimizes the operation of its interconnected elements from the central and distributed generator through the high-voltage transmission network and the distribution system, to industrial users and building automation systems, to energy storage installations and to end-use consumers and their thermostats, electric vehicles, appliances and other household devices.

The definition shows that the NIST has a focus on the transition process to a Smart Grid. In general, every Smart Grid definition comprises technical aspects and domains as well as actors. As the definitions show, one key element of the Smart Grid transition is the integration of new technologies from the ICT domain with the existing grown infrastructure of automation and power distribution and transmission. As legacy processes and systems have to be integrated with new functionality and components, from ICT perspective, classic enterprise application integration (EAI) projects arise with all of their requirements and problems. As systems are integrated, they have to agree on both a syntactical and semantical level on the concepts, profiles and serialization of the data to be exchanged between the systems. Therefore, a common information model is of highest importance.

1.2 General Motivation for Standardization in Smart Grids

The requirement for system openness as well as the necessary amount of data exchange between participating parties and components inside a Smart Grid ecosystem leads to standardization to fulfill the interoperability requirements in a holistic architecture and to enable a smart, ICT-driven transmission and distribution grid. Without standardization (e.g. in terms of data models and interfaces) costs for integrating components as well as applications would be enormous due to the large number of new interfaces and processes involved. The German national standardization strategy [2], for example, defines amongst others issues the following goals which can be achieved with the support of standards:

- Standardization can act as a strategic instrument to support both the economical and social success.
- Standardization relieves the regulatory activities of the government
- Standard Developing Organization (SDOs) foster technology convergence and advancement
- SDOs provide efficient processes and instruments for industrial participation and coordination

The various regional and international initiatives for Smart Grid standardization (see a summary from a project in [9] or [11][2]) outline the importance of standards in the Smart Grid domain. Most of them name and recommend single standards or families of standards, but they have several recommendation in common which are summarized in the next sub-section.

1.3 Internationally Recommended Core Standards for Communications and Data Modeling

Different countries, organizations and vendors came up with their own roadmaps regarding the Smart Grid standardization. Some countries focus on solutions at the customer's site which often have different focus from home appliances for demand response (DR), peak shaping, home automation, other countries mainly focus on reducing non-technical losses through smart metering or improving the outage management for radial feeder systems.

One main scope overall seems to be to provide the reliability of supply for the digital economy, focusing on markets and economic benefits for the country. Another strong point is to integrate more sustainable energy, trying to reduce carbon dioxide emissions and to cope with distributed, renewable generation like Micro Combined Heat and Power (CHP), photovoltaics, and fuel cells as well as with electric vehicles. A one-size-fits-all solution for those Smart Grid requirements both in terms of technical solutions and its corresponding standardization requirements is unlikely to be found.

[2] Both were also basis for [4] and [1].

The main intention of providing a table of standards in [4] is to facilitate an overview of international standards like IEC (International Electrotechnical Commission) or ISO (International Organization for Standardization) standards, so only few national standards have been integrated. Looking through the table, the most striking and common core standards are from the IEC TC 57 which will can be briefly seen in figure 1.2. A more detailed overview on the Seamless Integration Architecture (SIA) can be found in [12].

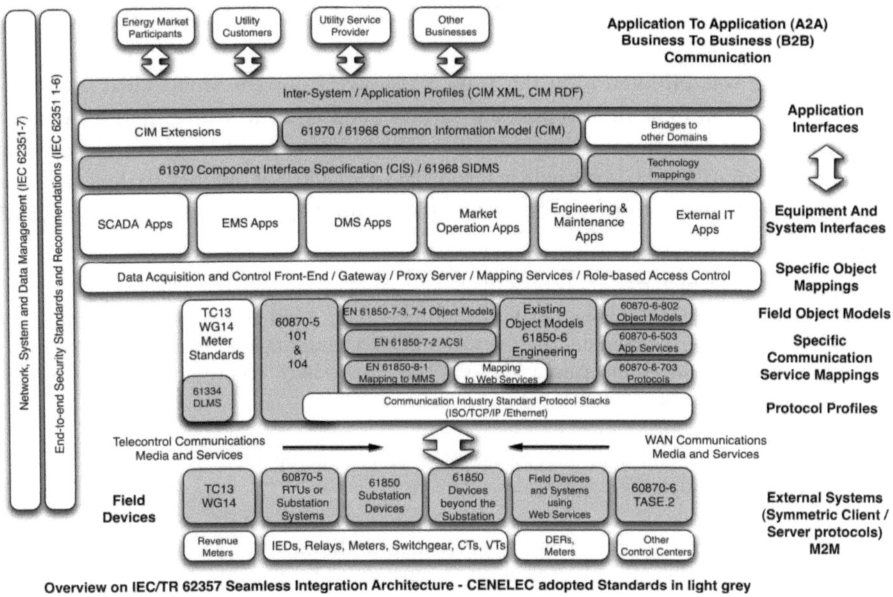

Fig. 1.2 The IEC 62357 Seamless Integration Architecture

According to the number of recommendations, the following TC 57 standards are of highest importance from the perspective of most experts: Since national perspectives differ, organizational standards, e.g., like Institute of Electrical and Electronics Engineers (IEEE) ones have less impact on worldwide scale.

- IEC TR 62357: Reference Architecture - Seamless Integration Architecture SIA
- IEC 61968/61970: Common Information Model (CIM) for Energy Management Systems EMS and Distribution Management Systems DMS
- IEC 61850: Intelligent Electronic Device (IED) Communications at Substation level and Distributed Energy Resources (DER)
- IEC 62351: Vertical Security for the TR 62357
- IEC 60870: Telecontrol protocols (though mostly expected to be a deprecated legacy standard)
- IEC 62541: OPC UA - OPC Unified Architecture, Automation Standard
- IEC 62325: Market Communications using CIM

Within this book we do not present the initiatives which have led to those conclusions (see [13] for a very detailed overview on those topics) but focus on the introduction and application of the aforementioned key standards except the 60870 standards family and the SIA in general. The application of those standards is bound to use cases and architectures, this book will provide a quick overview on the internal functioning of the standards and methods to be applied in context.

1.4 The EU Mandate M/490 to CEN/CENELEC and ETSI

The mandate was initiated after the Joint Working Group on Smart Grid standardization [3] stated in their final report the further need to cope with Smart Grid standardization and the european aspects of it. The original Joint Working Group was renamed to Smart Grid Coordination Group SG-CG and got the task to deal with the M/490 mandate and its aims, as described in the following excerpt from the original mandate in english [6]:

> The Smart Grid Task Force has identified very strong requirements for inter-operation of a large variety of domains (such as Grid operation, Grid automation, Distributed Energy resources management, Industry automation, Building and Home automation, Smart metering) while insuring a high level of consistency, security, data protection and privacy, and cost efficiency.
>
> All these domains and their integration into a single interoperable system are also at different steps in maturity.
>
> A secure and robust energy network is essential for the continuous improvement and industrious operation of the European energy markets. This will only be possible if the associated information and communication networks are secure and robust. It is also essential to maintain data and system security and to respect the rights of end consumers as well as the fundamental rights and freedoms of natural persons.
>
> As stated above, the scope of Smart Grids is large; thus the risk is that too many standardization bodies work on this issue, providing inconsistent sets of technical specifications, causing non-interoperability of equipment and applications and that the priorities will not be precisely defined.
>
> The challenge of Smart Grids deployment will require changes to existing standards, industry rules and processes.
>
> This mandate is to address such a challenge in the field of standardization. The expected long term duration of Smart Grid deployment suggests the need for a framework that is:
>
> - Comprehensive and integrated enough to embrace the whole variety of Smart Grid actors and ensure communications between them

[3] http://www.cenelec.eu/aboutcenelec/whatwedo/technologysectors/smartgrids.html

- In-depth enough to guarantee interoperability of Smart Grids from basic connectivity to complex distributed business applications, including a unified set of definitions so that all Members States have a common understanding of the various components of the Smart Grid.
- Flexible and fast enough to take advantage of the existing telecommunications infrastructure and services as well as the emergence of new technologies while enhancing competitiveness of the markets
- Flexible enough to accommodate some differences between EU Member States approaches to Smart Grids deployment

The value of such a framework will also be to foster and develop convergence of standards.

CEN, CENELEC, and ETSI are requested to develop a framework to enable European Standardization Organizations to perform continuous standard enhancement and development in the field of Smart Grids, while maintaining transverse consistency and promote continuous innovation. The expected framework will consist of the following deliverables:

1. A technical reference architecture, which will represent the functional information data flows between the main domains and integrate many systems and subsystems architectures.
2. A set of consistent standards, which will support the information exchange (communication protocols and data models) and the integration of all users into the electric system operation.
3. Sustainable standardization processes and collaborative tools to enable stakeholder interactions, to improve the two above and adapt them to new requirements based on gap analysis, while ensuring the fit to high level system constraints such as interoperability, security, and privacy, etc.

This framework will build on the Smart Grid Task Force reports from Expert Group 1 (EG1, especially on chapter 11), EG2 and EG3 as main inputs, as well as already existing material delivered through other mandates such as the M/441 and M/468.

Specifically regarding information security and data privacy, standards will be developed and enhanced in order to encompass an agreed set of harmonized high level requirements as proposed by the Smart Grid Task Force.

In response to the mandate, the Smart Grid Co-Ordination Group was formed on 01.07.2011. This group emerged from the earlier Joint Working Group on Standards for Smart Grids, which produced a report on European Standardization of Smart Grids from 01.06.2010 to 31.12.2010. Its original organization structure is shown in figure 1.3.

The Smart Grid Co-ordination Group is not a standardization body (i.e. A technical committee (TC) of a standards defining organization (SDO)) as such, but rather a co-ordination group to steer and support the execution of the mandate (details can be found in their Terms of Reference (ToR). The Smart Grid Co-Ordination Group established four working groups corresponding to the deliverables of the mandate:

- Working Group First Set of Standards WG FSS
- Working Group Reference Architecture WG RA

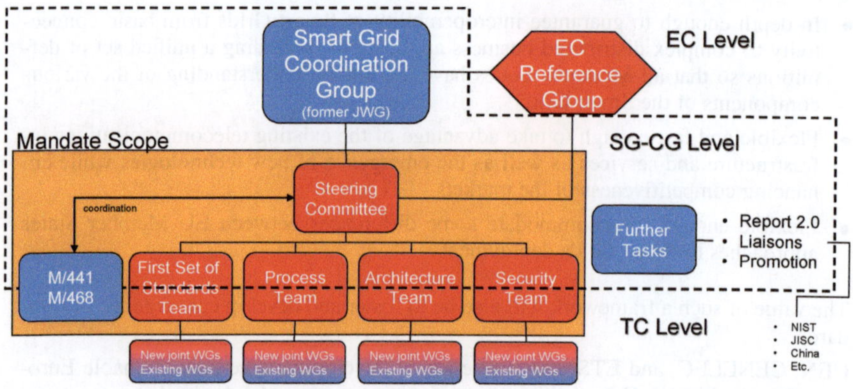

Fig. 1.3 The structure of the SG-CG

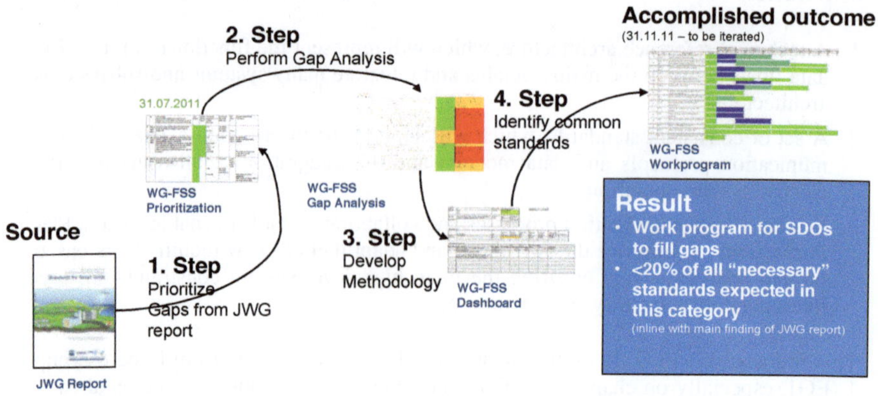

Fig. 1.4 Approach to find gaps in the M 490 mandate from the FSS group

- Working Group Sustainable Processes WG SP
- Working Group Smart Grid Information Security WG SGIS

Furthermore, close relations were established with the mandate M/441 (Smart Metering) and mandate M/468 (E-Mobility) through so called rapporteurs. Within the mandate, the four groups have to provide deliverables in form of reports which will be given to standardization being mandatory like a technical report from IEC. As the process of the groups started in parallel, the normal way to do everything in a cycle could not be established. Therefore, starting with use cases, coming up with an architecture based on those functions derived form the use cases and then defining standards for the functions in the architecture and identify new needed standards was not possible. Since work had to be done in parallel, inconsistencies in the work deliverables might occur (wording, glossary, technologies).

Figure 1.4 shows the future process which has to be taken to deal with the existing work from the various ETP and JWG groups on Smart Grid while figure 1.5 depicts the existing work from the four teams and its interactions.

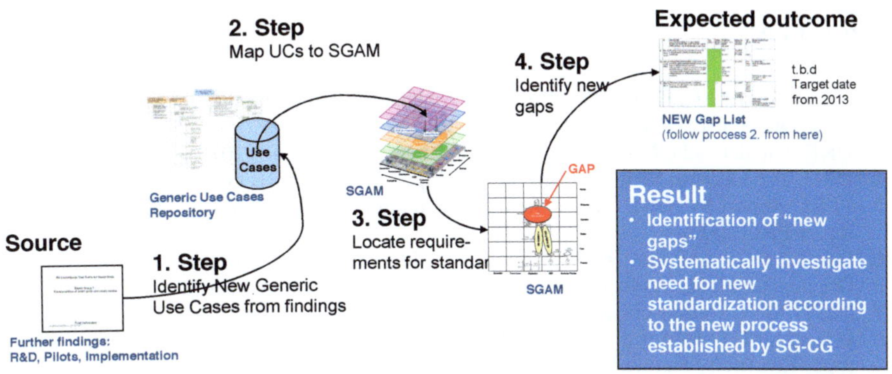

Fig. 1.5 Approach to find gaps in the M 490 mandate for future work from the FSS Group

1.5 Conclusion

Smart Grid is clearly an emerging topic which has been well established over the last years. However, the focus of Smart Grid depends on the region where it is applied—it can be either economics (power theft, distribution outage management, remote metering and billing, grid extensions costs), ecology (carbon-dioxide reduction, integrating renewables) or security of supply (efficient use of assets—the power grid, integration of DER, VAr or frequency regulation) or basically, a weighted a mixture of those three dimensions. This is also reflected by the various definitions which have been presented in this chapter.

The focus for the remaining chapters of this book is on the view from the European perspective from the EU mandate M/490. The mandate brought together experts from the most relevant and important European SDOs which could gather to harmonize the needed standards and technologies for Smart Grid as well as agree on architectures and future needed technologies (respectively standards).

Within this book, we will provide short introductions to the most striking standards regarding the communication and ICT aspects of the European Smart Grid. One of the aspects is architecture management, with a strong focus on how to meaningfully cope with enterprise architectures in a utility, adopting the Smart Grid ArchitecturalModel (SGAM) for the M/490 RAWG for proper use case management, applying the main recommended standards for Smart Grid, IEC 61850 and CIM to operation, harmonizing those two standards and deal with semantic data models in a utility and an outlook on future applications like electric vehicles, seamless market communications and future automation using OPC UA. The individual chapters can be read as basic building blocks, being self contained with individual references and further readings but can also be seen in the light of the bigger picture from the M/490 perspective.

References

1. CEN, CENELEC, ETSI: JWG Report on Standards for the Smart Grid. Tech. rep. (2010)
2. DIN: Die deutsche Normungsstrategie aktuell (2009)
3. DKE: Die deutsche Normungsroadmap E-Energy/Smart Grid (2010)
4. DKE: The German Standardization Roadmap E-Energy/Smart Grid. VDE (2010)
5. EPRI: Report to NIST on the Smart Grid Interoperability Standards Roadmap. Tech. rep., EPRI (2009)
6. European Commission: M/490 Standardization Mandate to European Standardisation Organisations (ESOs) to support European Smart Grid deployment (2011)
7. NIST: NIST Framework and Roadmap for Smart Grid Interoperability Standards (2010)
8. Ø stergaard, J.: European SmartGrids Technology Platform-Vision and Strategy for Europes Electricity Networks of the Future (2006)
9. Rohjans, S., Uslar, M., Bleiker, R., González, J.M., Specht, M., Suding, T., Weidelt, T.: Survey of Smart Grid Standardization Studies and Recommendations. In: First IEEE International Conference on Smart Grid Communications (2010)
10. SMB Smart Grid Strategic Group (SG3): IEC Smart Grid Standardization Roadmap (2010)
11. Uslar, M., Rohjans, S., Bleiker, R., González, J.M., Suding, T., Specht, M., Weidelt, T.: Survey of Smart Grid Standardization Studies and Recommendations - Part 2. In: IEEE Innovative Smart Grid Technologies Europe (2010)
12. Uslar, M., Rohjans, S., González, J., Specht, M., Trefke, J.: Das Standardisierungsumfeld im Smart Grid - Roadmap und Outlook. Elektrotechnik & Informationstechnik 128(4) (2011)
13. Uslar, M., Specht, M., Rohjans, S., Trefke, J., González, J.M.: The Common Information Model CIM: IEC 61968/61970 and 62325 - A Practical Introduction to the CIM. Springer (2012)

Part II
Requirements and Architectures

Chapter 2
Requirements Engineering for Smart Grids

Christian Dänekas and José M. González

Abstract. Within this chapter requirements engineering methods and models in context of Smart Grid development are outlined. First, the activities of requirements engineering and their integration into superordinate process models are introduced. Subsequently, the layered models of the GridWise Architectural Council and the Smart Grid Architectural Model (SGAM) are outlined. These cover viewpoints from business requirements to component specification and therefore may be used to structure Smart-Grid-specific requirements engineering activities. The application of the SGAM is illustrated by using the example of the Advanced Metering Infrastructure (AMI). Since the development of Smart Grids represents a long-term engineering effort, this chapter also introduces the concept of technology roadmaps to manage strategic goals and requirements. The assessment of technological and organizational progress in Smart Grid development complements this approach and is subsequently covered by the discussion of the Smart Grid Maturity Model (SGMM). The chapter concludes with a brief summary and references for further reading regarding inputs to the SGAM Layers.

2.1 Motivation

The concept of Smart Grids shall enable the generation and consumption of electrical power to become more efficient and sustainable to meet the challenges of climate change and reduce the dependence on fossil fuels and nuclear power. To achieve this, the coordination between distributed generation assets, storages and the consumers shall be realized by using Information and Communication Technologies (ICT). A common definition adheres to this idea by defining Smart Grid as an "electricity network that can intelligently integrate the actions of all the users

Christian Dänekas · José M. González
OFFIS – Institute for Information Technology, Escherweg 2, 26121 Oldenburg, Germany
e-mail: {daenekas,gonzalez}@offis.de

M. Uslar et al.: *Standardization in Smart Grids*, POWSYS, pp. 15–37.
DOI: 10.1007/978-3-642-34916-4_2

connected to it—generators, consumers and those that do both, in order to efficiently deliver sustainable economic and secure electricity supply" [6].

The scale of the existing infrastructure and its criticality regarding economical growth may induce high costs, if its further development towards a Smart Grid lacks transparency. Smart Grid pilots and projects are continuously developing new business concepts and possible technical scenarios. In order to compare and benchmark different concepts regarding their strengths and weaknesses, it is necessary to classify and document the functional and non-functional requirements connected to them. This also addresses interoperability between power system components as a key factor of a Smart Grid. Standardization may support the goal of interoperability while maintaining the opportunity for innovation. In order for Smart Grid development to succeed, the requirements which serve as the system's basis shall be engineered with the same amount of care as its physical and informational infrastructure. This chapter aims at providing an overview regarding requirements engineering concepts and methodologies and illustrating their application at different levels of the Smart Grid engineering process. This covers issues which are subject to standardization as well as aspects highly relevant within the business/enterprise context.

First requirements engineering concepts and their process integration will be introduced (Section 2.2), followed by the description of structural models supporting Smart Grid requirements engineering (Sections 2.3 and 2.4). The application of the European Smart Grid Architectural Model (SGAM) representing such a structure is outlined regarding the technological field Advanced Metering Infrastructure (Section 2.5). In order to relate specific Smart Grid engineering and standardization projects to the long-term system engineering process, technology roadmaps and maturity models are introduced as means to cope with the management of long-term Smart Grid requirements (Sections 2.6).

2.2 Requirements Engineering Concepts and Process Integration

The Smart Grid as a system exhibits a high complexity regarding organizational and technological aspects. Various actors take part in the planning and construction of the system representing several organizations and engineering domains [1]. Therefore, a key challenge of the Smart Grid is integration, affecting components for generation, transportation, distribution, storage, and consumption of electrical energy and the supporting information systems and applications. To create the Smart Grid as an operational system-of-systems, the functionalities and interfaces of its artifacts must be specified beforehand. As requirements serve as the decisive factor for all further engineering activities, a suitable methodology for requirements specification and management is essential. This ensures traceability between design decisions and the system requirements, supports the collaboration between stakeholders by assigning responsibilities, allows to derive the structure of the system regarding software and hardware artifacts and enables to test the implementation against the specification.

Requirements Engineering as discussed in this chapter stems from the software engineering discipline. Software engineering itself, including the elicitation and management of requirements was established as a possible solution to the "software crisis" in the 1970s. Around that time software development was becoming more complex as a result of strong increase of computing power [7]. The term "Requirements Engineering and Management" was explicitly introduced in the 1990s, referring to the elicitation, management, and analysis of requirements [7]. In literature and practical experience the terms Requirements Engineering, Requirements Management or Requirements Analysis are sometimes used interchangeably. In other cases different terms are used to emphasize a certain aspect. This chapter uses Requirements Engineering as the main term to refer to the activities of elicitation, analysis, negotiation, documentation, validation, and management of requirements. These activities are depicted in Figure 2.1 and shall be defined as follows (see [11]):

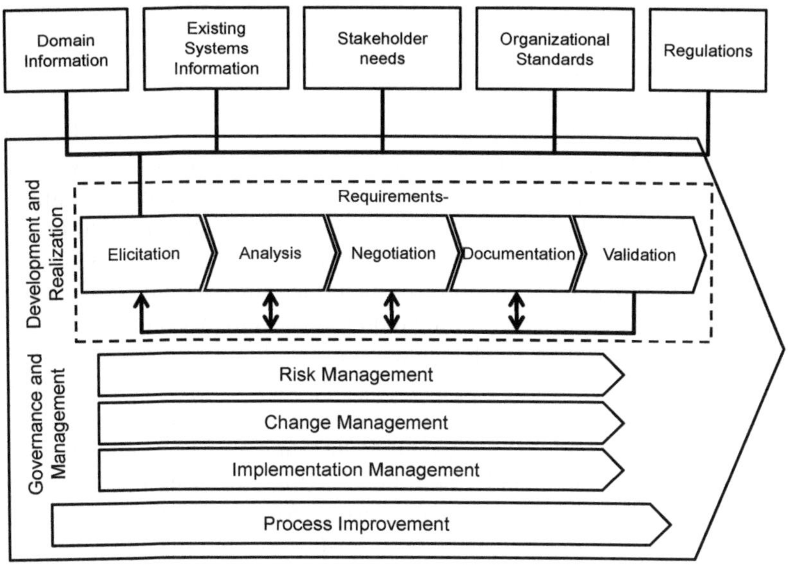

Fig. 2.1 Requirements Engineering activities and process inputs based on [17] and [11]

- ***Elicitation:*** The initial set of system requirements is derived. This is done in collaboration with the systems' stakeholders. Among others documentation regarding domain knowledge in general, relevant and mandatory (organizational) standards and regulations, specifications of other systems (e.g. legacy or associated systems) should also be reviewed in this step.
- ***Analysis and negotiation:*** The initial set of requirements is reviewed in collaboration with stakeholders. Conflicts between the requirements themselves and

with time and budget constraints of the project shall be resolved resulting in a list of accepted requirements.

- ***Documentation:*** The requirements negotiated with the systems' stakeholders are documented to an appropriate level of detail. This may be done using natural language, appropriate templates, modeling or formal specification, depending on the tools available and the aspect of the system specified by the requirement in question.
- ***Validation:*** The requirements shall be continuously reviewed regarding consistency and completeness.
- ***Management:*** The management of requirements shall help to establish traceability regarding changes to the requirements and their dependencies on each other. This activity is of high importance through the whole system life-cycle and is conducted in parallel to the activities above.

The requirements management process consists of several subprocesses ([17]). Within an engineering project, for example the number of requirements or requirement changes might strongly increase representing a risk which might lead to the projects failure. *Risk Management* therefore comprises activities to identify such risks, estimating their probability and severity, and providing measures to handle or mitigate them. The change of requirements is further addressed within the *Change Management* process which defines how changes to requirements shall be handled. This includes the definition of a change request template, creation of an impact analysis process and management of requirement configurations. Finally the *Implementation Management* process shall coordinate the subprocesses of requirements engineering and establish traceability from requirement specification to implementation. The optimization of the development and management processes is addressed within *Process Improvement*. The improvement measures should be aligned to the organizations current maturity regarding requirements engineering.

As requirements and their implementation mutually depend on each other, Requirements Engineering is not a standalone approach. In Software Engineering therefore multiple methodologies integrate requirements engineering activities. In case of the waterfall model [16] for example, the elicitation of requirements represents the first phase of the model. The following phase of design uses the requirements specification to derive the systems design. Afterwards, the same procedure is applied to the phases of implementation, verification and maintenance of a system. Therefore the next phase shall only be begun if the preceding phase is completed. Such an approach however limits the capability to react towards changes during the requirements process, since the model proposes only one phase should be worked on at a time. If the verification of the systems leads to changes regarding the requirements the project must track back to the requirements phase. Depending on the project scale regarding time and budget, changing requirements can lead to failure of the project. The fact that requirements may change during a project is addressed by iterative models like the spiral model [2] or the Rational Unified Process (RUP) [12]. In contrast to the waterfall model, the project results including the requirements specification are continuously evaluated against each other using iterations. The RUP furthermore promotes modeling of requirements and integrating those models with the systems' design. Requirements shall be specified by Unified Modeling Language (UML) use cases combined with the application of suitable templates for requirements documentation[1].

To conclude, the integration of Requirements Engineering within engineering methodologies relates requirements to the systems stakeholders, project constraints regarding time and costs, implementation artifacts and test results. The assignment of roles and the specification of interfaces between stakeholders and the outcome of the processes phases lead to higher transparency regarding the engineering artifacts and project organization. Depending on the project context, different approaches may be used to achieve this, ranging from lightweight, agile methods to complex Enterprise Architecture Management (EAM) frameworks like The Open Group Architecture Framework (TOGAF) [20][2].

Furthermore, support is needed to deal with the complexity of the requirements themselves, e.g. by using fitting templates and models in order to speed a coherent requirements description and provide a common semantic. This helps engineers and other stakeholders belonging to different disciplines to understand the specification of the system and to collaborate with each other. The next section will outline a common approach to structure the Smart Grid regarding interoperability aspects. By using such a model, stakeholders are able to categorize requirements regarding the concerns at levels spanning from economic and regulatory policy to the basic connectivity of system artifacts.

2.3 Managing Smart Grid Requirements Regarding Interoperability Aspects

The development of the Smart Grid is expected to be a long-term transition. Since the power system itself is already in place, multiple actors or stakeholders are concerned with operation, maintenance, and business aspects of the power system. Also a great number of components and applications, needed to generate, transport and distribute electrical energy, is in place. Therefore, the Smart Grid as a system cannot be engineered from the ground up. Instead, Smart Grid development shall be characterized as a migration process. This means business models and market roles on the one hand, and technical components and architectural structures on the other hand, shall be migrated from the current "legacy" state towards the "Smart Grid". Due to the scale of the system and its economical importance, failures in operation and especially architectural and functional planning of the system potentially induce high costs. In order to enable a well structured migration process the requirements for the Smart Grid and the current system have to be decomposed using an appropriate model. Following the definition given in Section 2.1, interoperability represents an essential requirement for the Smart Grid, since it is supposed to integrate dif-

[1] A methodology for Smart Grid Use Case Management is outlined in Chapter 3.

[2] TOGAF as a framework for Enterprise Architecture Management is explained in detail in Chapter 4.

ferent assets and applications into one functional system. In order to support the elicitation and management of requirements, a suitable structure should be used. The GridWise Architectural Council (GWAC) accordingly proposes eight layers of interoperability, as shown in Figure 2.2.

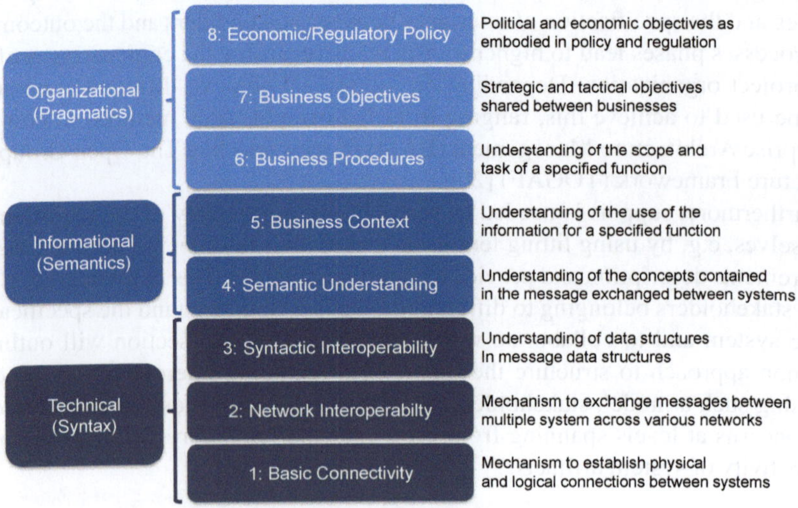

Fig. 2.2 Levels of interoperability defined by the Grid Wise Architectural Council (adapted from [18])

Starting top-down, the first three layers of the model commonly referenced as the "GWAC stack" are concerned with organizational aspects of interoperability. At the highest level (8), Smart Grid development is influenced to a high degree by *economic and regulatory policy*. Stakeholders of the system must adhere to rules and regulations determined by governments and regulators. The next level (7) is concerned with high-level collaboration between businesses concerning strategic and tactical *business objectives*. This is further elaborated by the alignment of *business processes and procedures* (6) by defining high-level interfaces consistent with the framework proposed at the higher layers. This helps the organization to further elaborate business functions and identifying which parts of these functions lie outside the organizations boundaries.

The next two layers build on the organizational aspects by transferring them into appropriate information models. To do so, first the *business context* has to be elaborated by identifying the need for information connected to the business functions (5). Depending on the functions considered, existing information models from different domains may be used as the foundation for more problem-specific models. After identifying the information needed to support the business functions embedded in processes and procedures, a common understanding regarding the semantics within the information models has to be established (4). This *semantic understanding* regarding the information contained in the models is especially important since the Smart Grid aims at the "plug and automate" paradigm, which means that technical assets and applications should be able to "understand" each other without a high amount of configuration by domain experts. Instead, the human knowledge on the meaning and rules applying to model entities has to be codified explicitly within the model itself.

The last three layers of the model deal with the technical interoperability aspects of the Smart Grid. While layer (4) dealt with the semantics of data model contents the next layer expresses the need for *syntactic interoperability* concerning the messages which are exchanged between applications and assets in the Smart Grid (3). The GWAC compares this layer to the seven layer ISO/OSI model [10], stating this layer subsumes its application and presentation layers, including character translation, content structure and message exchange patterns. Standards relevant to this layer, among others, include HTML, XML and SOAP. In order to enable the exchange of messages within the Smart Grid, *network interoperability* has to be present which is in focus of the next layer (2). With reference to the ISO/OSI model this layer addresses the network, transport, session and in parts the application layer. This includes functions like resolving logical into physical addresses or management of the exchange of data messages. Communication standards relevant to this layer, among others, include TCP, UDP, FTP, and IP(v6). Apart from syntactic and network interoperability, the technical part of the GWAC interoperability model covers the *basic connectivity* needed to establish networked communication between distributed assets and applications within its lowest layer (1). To complete the ISO/OSI model this layer addresses its physical and data link layers including functions like electrical connectivity, character encoding, data flow control or error correction. Relevant standards concerning this layer are among others Ethernet, WiFi or PPP (Point-to-Point Protocol).

The GWAC stack may serve as suitable structure for the elicitation and management of Smart Grid requirements, since it addresses the facets of Smart Grid interoperability from regulatory policy down to the physical connection between assets. Each layer is distinctive regarding the functions that are covered and the abstraction level or degree of formalization of the requirements connected to them. The authors of the model state that it may need context specific tailoring to fit the organization or generally speaking the context it shall be applied to. An important adaption of the model is currently done in context of the European Smart Grid Mandate M/490 (see [3]), strictly speaking as one of the basic concepts of the Reference Architecture developed in this context. The main output of this adaption is the European Smart Grid Architectural Model (SGAM), which is described in the next section.

2.4 Architectural Viewpoints towards Smart Grid Requirements

In the context of the European Commission's Standardization Mandate M/490, a holistic viewpoint of an overall architecture named Smart Grid Architecture Model

(SGAM) [3] (see Fig. 2.3) is developed. This work is based on existing approaches (like [8], [19] or [13]) and subsumes the different perspectives and methodologies regarding the conceptualization of Smart Grids. In Figure 2.3 the SGAM structure with its layers, domains and zones is outlined. SGAM comprises three core viewpoints layers, domains and zones which support a holistic view on architecture, see Figure 2.3.

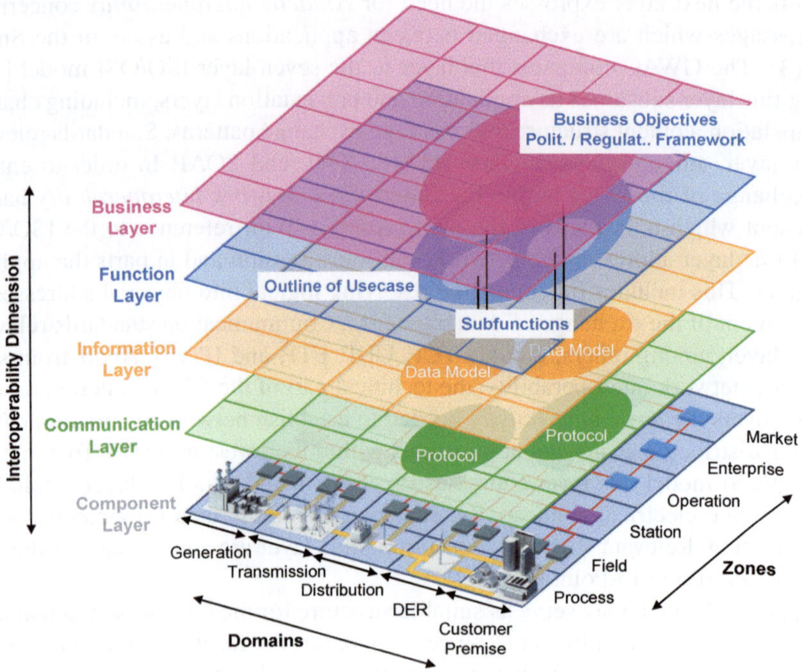

Fig. 2.3 SGAM Overview [3]

The layers of the SGAM are explained by [3] as follows:

- ***Business Layer*** Business viewpoint regarding strategic and tactical goals and processes as well as regulatory aspects.
- ***Function Layer*** IT-oriented, technology independent description of use cases, functions and services.
- ***Information Layer*** Business objects and data models of the Function Layer to enable interoperability.
- ***Communication Layer*** Specification of protocols and procedures for the data exchange between components based on the Information Layer.
- ***Component Layer*** Physical and technical view on Smart Grids components. Besides power-system related infrastructure and equipment, ICT-infrastructure and -systems are also considered.

These layers were adopted by the GWAC stack. Therefore the eight layers of the GWAC stack can be mapped onto the five distinctive layers of the SGAM like depicted in Figure 2.4.

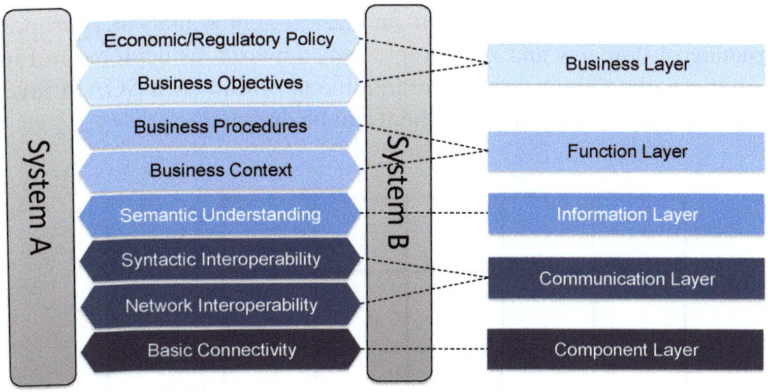

Fig. 2.4 Adoption of the GWAC Stack in context of the European Smart Grid Reference Architecture (adapted from [3])

The aim of the adoption of the GWAC model within the SGAM is to improve conversation regarding architectural requirements for the Smart Grid. It should enable the elicitation of functional and non-functional requirements in context of each individual layer and the interdependencies between them. Misunderstandings can be reduced by explicitly naming the level a certain requirement addresses.

In addition to the adoption of the GWAC layers, the SGAM proposes a structure of domains and zones. On each layer, the horizontal dimension on the one hand comprises domains of the energy industry: generation, transmission, distribution, distributed energy resources and customer premise. On the other hand, the vertical dimension describes a hierarchical structure of different zones: market, enterprise, operation, station, field and process.

In summary, the SGAM provides through its three viewpoints—layers, domains and zones—a generic technology neutral view on Smart Grids which can illustrate various power system architectures. Therefore, SGAM and its three interrelated viewpoints are used in Section 2.5 to exemplarily illustrate how requirements can be gathered regarding an Advanced Metering Infrastructure.

2.5 Exemplary Requirements Analysis for Advanced Metering Infrastructure Using the Smart Grid Architectural Model

Several Smart Grid functionalities like Demand Side Management or dynamic (load and time variable) tariffs require the provision of metering data in a fast and aggregated way. Advanced Metering Infrastructure[3] (AMI) aims at providing infrastructure as well as hard- and software to operate and manage digital meters [9]. Consequently, AMI is part of several advanced Smart Grid services and shall be used within this section to illustrate a SGAM-based requirements analysis.

The SGAM, introduced in the previous section, shall ensure a holistic analysis of requirements. For each layer, a domain- and zone-oriented analysis is proposed here. The structure of domains and zones represents a matrix, as depicted in Figure 2.5. Based on these two dimensions, the requirements within each SGAM layer can be classified and relationships between them can be outlined.

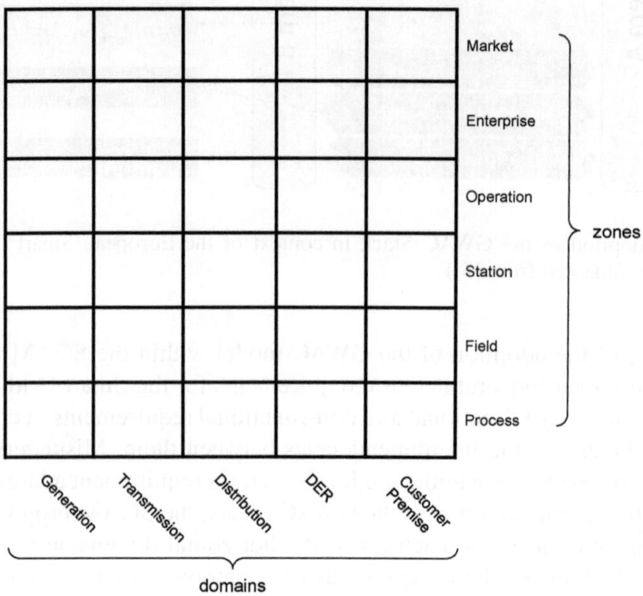

Fig. 2.5 SGAM domains and zones based on [3]

The application of this matrix for each SGAM layer—Business, Function, Information, Communication and Component—shall help to handle the complexity of Smart-Grid-related architectures. In the following sections, requirements regarding the technology field AMI with respect to the SGAM layers are outlined.

2.5.1 Business Layer

Regarding the Business layer the goals of AMI can be subsumed simplified by three main aspects, depicted in Figure 2.6 as Advanced Functionality, Smart Metering

[3] Sometimes also referenced as Automated Metering Infrastructure.

and Metering Services. The business functions related to these goals build on each other:

- **Metering Services:** Comprises basic functionality of metering devices regarding reading of metering data (like for example multi-utility support) and corresponding interfaces for remote reading.
- **Smart Metering:** Extends Metering Services by providing further functions which enable to gather metering data for, e.g., monthly billing and related services like provision of aggregated or detailed metering data.
- **Advanced Functionality:** Builds on Smart Metering and for instance supports dynamic tariffs and Demand Management. Therefore the functions underlying Smart Metering have to be available.

Figure 2.6 shows on the one hand Business goals classified according to the SGAM domains and zones matrix (left side). On the other hand guidelines and laws which affect these goals (*(1)* to *(3)*) are exemplarily listed. As highlighted in Figure 2.6 only the domains Distribution, DER and Customer Premise are addressed.

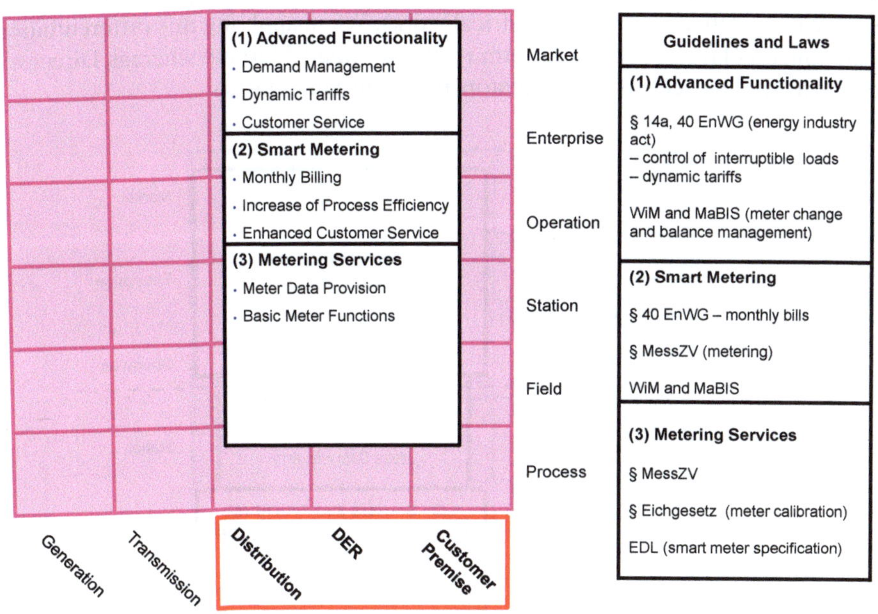

Fig. 2.6 AMI Business Layer Example

2.5.2 Function Layer

The function layer focuses on the IT-related functions which support corresponding business functions (business functions are printed in italics in Figure 2.7). Basically three function blocks are presented: *Service Platform*, *Directory Service* and *Sensors and Actuators* for DER, see Figure 2.7 *(1)* to *(3)*.

- **Service Platform:** Comprises AMI service platforms for different market roles like distribution network operator, DER operator or customer. Here, meter data is prepared according to the requirements of each market role.
- **Directory Service:** Includes functionality to discover and control metering devices.
- **Sensors and Actuators Gateways:** Addresses several Gateways and Interfaces to access and control sensors and actuators regarding metering data. Here the focus is primarily on network stability.

Within its position paper [4] the German BNetzA differentiates between *Smart Grids* and *Smart Markets*. Smart Grids in this context refer to the serving role of the distribution system operator focusing on the provision of energy capacities (kW – power). In contrast, Smart Markets address the energy market perspective focusing on energy delivered (kWh – electrical work). When applying this differentiation to the function layer, Service Platform relates to Smart Markets whereas Directory Service as well as Sensors and Actuators relates to Smart Grids.

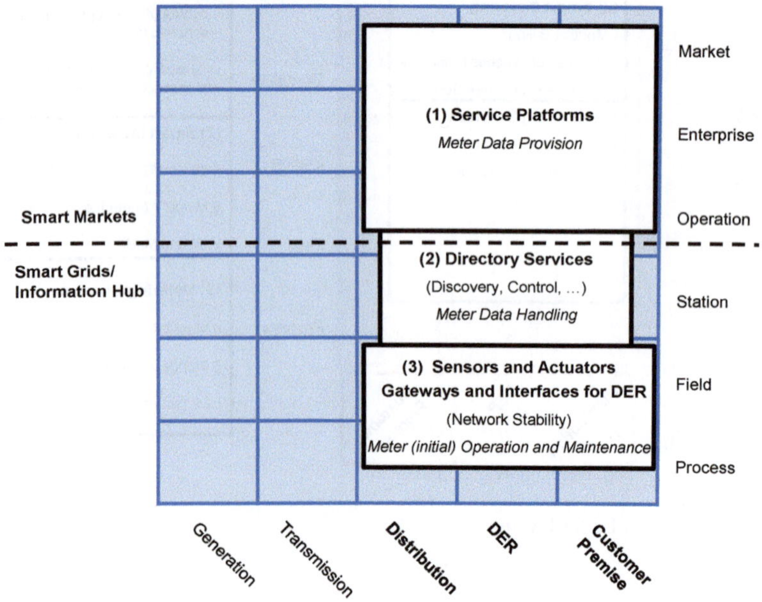

Fig. 2.7 AMI IT Function Layer Example

2.5.3 Information Layer

The IEC Technical Committee 57 "Power Systems Management" and associated information exchange elaborated within its working groups several standards for a seamless integration of market participants, applications and field devices. These standards are part of the corresponding TC 57 Reference Architecture referenced as the IEC Seamless Integration Architecture (SIA) [8]. Within the SIA the IEC Common Information Model (CIM) serves as the abstract information model for all entities in the energy market. For an introduction to the CIM see Chapter 6.

Based on the SIA, standards concerning the information layer of AMI were chosen exemplarily, like depicted in Figure 2.8[4]. They may be divided into three groups (depicted as *(1)* to *(3)*). The first group contains market-related IEC standards (like IEC 61970) as well as German regulatory data formats (EDIFACT). The second focuses on standards relevant for the SGAM zones Enterprise and Operation and enhances the standards of group *(1)* by including, e.g. an ANSI standard for revenue metering. Finally, the third group includes standards related to the integration and control of metering devices in the field.

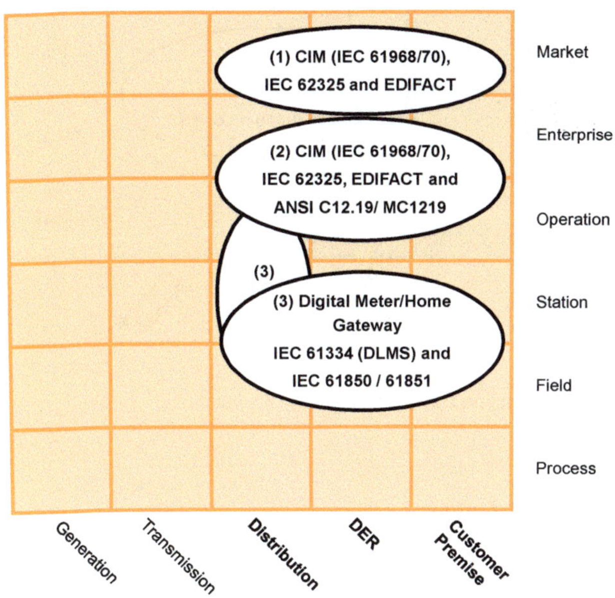

Fig. 2.8 AMI Information Layer Example

[4] The listed standards should not be considered exhaustive.

2.5.4 Communication Layer

Following the proposed standards by the IEC SIA outlined in Section 2.5.3, Figure 2.9 lists protocols possibly relevant for an AMI solution. These protocols may be basically divided into two groups according to the SGAM zones addressed. The first group covers use cases regarding market and enterprise data exchange based on the IEC 61968 Part 9 Meter Reading and Control. The second group comprises protocols which focus on the SGAM zones Operation, Station and Field (e.g. GOOSE or Zigbee).

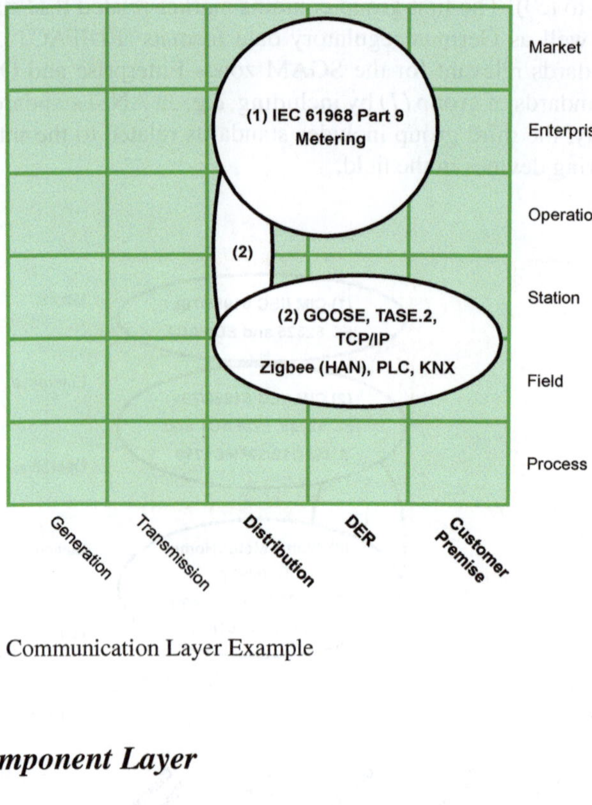

Fig. 2.9 AMI Communication Layer Example

2.5.5 Component Layer

The Component layer represents the lowest layer of the SGAM. Regarding AMI, it basically includes two types of components. On the one hand, IT-applications (e.g. commercial applications like billing) implement the aspects of the previous layers. On the other hand, technical equipment integrates information and communication technologies (e.g. Substation). Exemplary AMI-related components are depicted in Figure 2.10 numbered from *(1)* to *(11)*. In this context, core AMI IT-applications and equipment (residing in the dashed box like operational applications) and surrounding ones like commercial applications or transformers and network equipment can be differentiated.

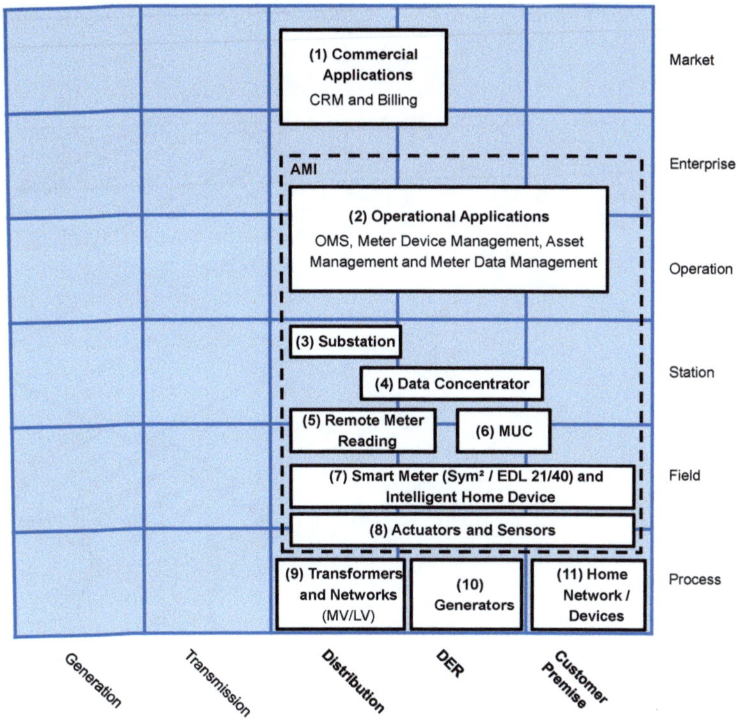

Fig. 2.10 AMI Component Layer Example

2.5.6 *Overview*

To summarize the exemplary AMI SGAM application, the five layers shall be analyzed in context of each other. Figure 2.11 depicts the overview on the results obtained in the previous sections. An analysis of the AMI architecture elements regarding the coverage of the SGAM's domains and zones across the five layers is shown in Figure 2.12. The coverage is divided into four levels: Core, High, Medium and Low.

Due to the duties and responsibilities of distribution network operators regarding AMI within all SGAM layers, the domain *Distribution* is nearly completely marked as an AMI core area (except for the *Process* zone). In addition, the zones *Operation* to *Field* for the domains *Distribution* to *Customer Premise* are mainly marked as core as well. The zones *Market* and *Enterprise* regarding the *DER*-domain and *Operation* concerning the *Customer-Premise*-domain are marked as high. As *Customer-Premise*-oriented services only build upon the core AMI functionality the related *Market* and *Enterprise* zones are rated as medium. Finally, as AMI primarily aims at the management and provision of meter data, technical equipment is only supervised. Therefore the *Process* zone is marked as low.

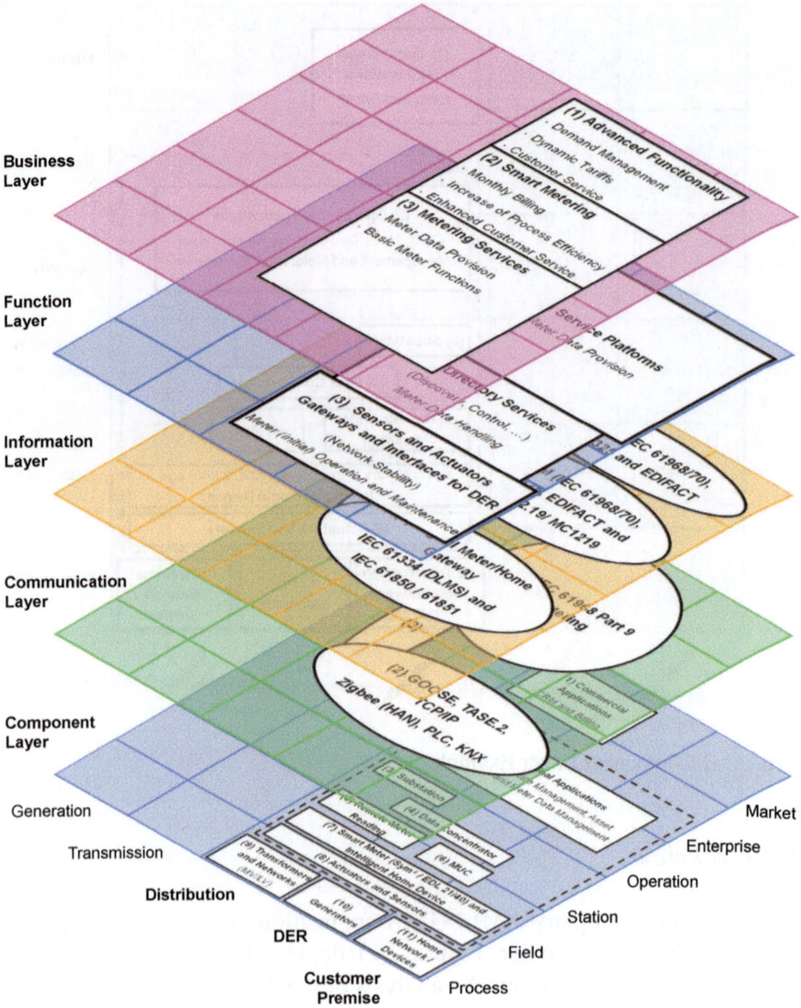

Fig. 2.11 Overview on the SGAM layer application regarding AMI

For further elaboration of the requirements analysis for AMI and other technology fields of the Smart Grid, the consideration of additional information sources is recommended. The dependencies of those fields should be carefully outlined, as their implementation and deployment is expected to be a long-term effort and many functionalities require the collaboration of various sub-systems. In the next section accordingly technology roadmaps and maturity models shall be introduced to support the management of long-term Smart Grid requirements.

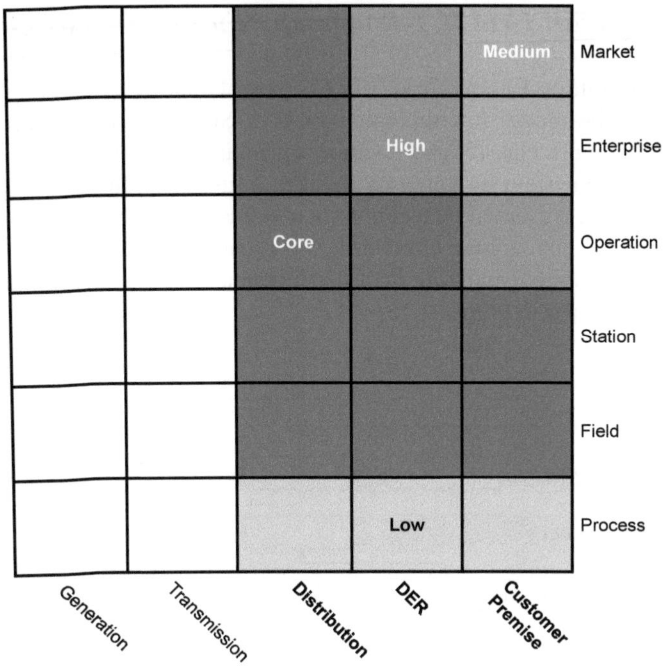

Fig. 2.12 Example AMI coverage regarding SGAM zones and domains

2.6 Management of Long-Term Smart Grid Requirements

As stated in Section 2.1, the Smart Grid represents a system of high complexity involving multiple stakeholders and a high amount of uncertainty regarding the implementation of the system. The time frame for the migration from the current power system to the Smart Grid is therefore expected to be decades rather than years. The overall development process should accordingly be considered in context of requirements engineering, complementary to the in detail specification of specific Smart Grid components. Roadmaps and maturity models may be used to provide high-level requirements covering the business and functional perspective towards the Smart Grid (see Figures 2.2 and 2.4). The top-level view of these models this way serves as valuable input for structures like the SGAM, like shown in Section 2.5 for the technological field AMI. Enterprises may adopt these models to design ICT-Architectures supporting their Smart Grid applications and services (see [23]). In the following, the technology-oriented roadmap resulting from the project "Future Energy Grid" [1] and the Smart Grid Maturity Model (SGMM) [21] will exemplify the structure and content useful for long-term Requirements Management.

2.6.1 The Smart Grid ICT-Roadmap "Future Energy Grid"

In the project "Future Energy Grid" (FEG) (see [1] and [5]), a roadmap was created regarding the means for implementation of Smart Grids in Germany till the year 2030. Roadmaps classify and document primary functional and non-functional requirements of a system by applying them to a time line. In case of FEG, the requirements were represented by technological fields which are decomposed into development steps. The roadmap afterwards was created by applying the development steps to a timeline and visualizing their dependencies among each other resulting in the structure depicted in Figure 2.13.

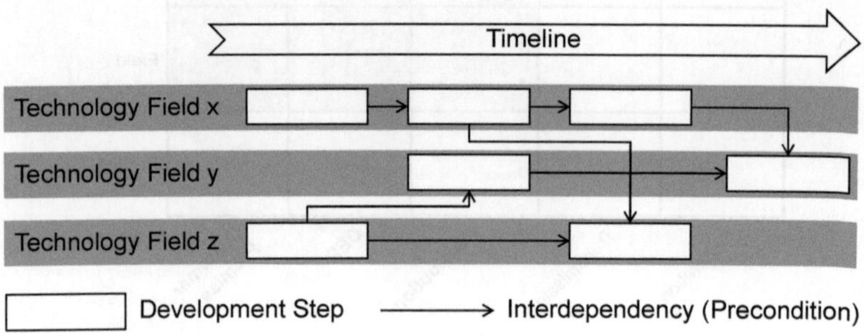

Fig. 2.13 Abstract structure of a technology roadmap

In order to create the FEG-roadmap, which is described in detail within the resulting study [1], the following steps were conducted:

1. Identification of key factors for the development of Smart Grids
2. Creation of scenarios by using the key factors projections
3. Selection of essential ICT-oriented technological fields for Smart Grid implementation
4. Elicitation of future development steps for the technological fields
5. Modeling the interdependencies between the development steps

The research approach chosen in FEG relates the technological options expressed by the development steps and the influential factors like the expected the energy mix or political and regulatory decisions expressed by the scenarios. As the results were worked out in close cooperation with experts from industry and research they cover a broad spectrum of viewpoints towards the Smart Grid. Also, the technological fields providing the basis of the roadmap were adopted within the functional viewpoint of the European Smart Grid Reference Architecture [3] as depicted in Figure 2.14. Therein, they are allocated to the domains of the Smart Grid system. Because the roadmap relates different viewpoints towards the Smart Grid within a long-term engineering process, it may be used to derive and evaluate requirements at the strategic level within an enterprise (see [23] regarding current research on this

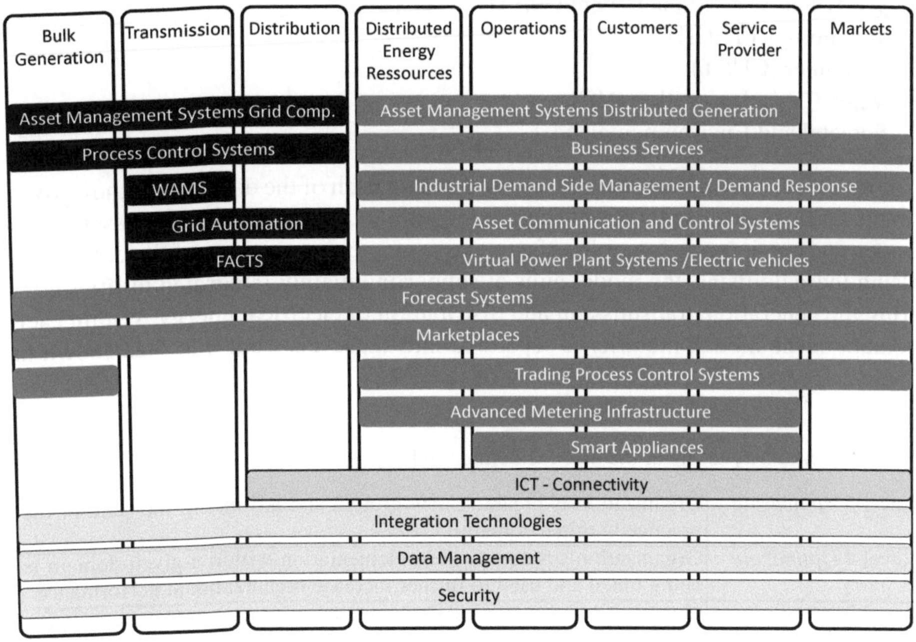

Fig. 2.14 Functional viewpoint of the European Smart Grid Reference Architecture [3]

issue). Further management assistance at this level is provided by the concept of maturity models. This will be discussed in the next section.

2.6.2 The Smart Grid Maturity Model

Maturity models are strongly connected to technology roadmaps as they aim at assessing the progress made regarding the goals connected to the roadmap. This way different Smart Grid strategies may be compared and benchmarked. The development of maturity models addressing enterprises and their progress regarding Smart Grid implementation is still a quite novel approach within the power industry. However, the Smart Grid Maturity Model (SGMM) [21] stewarded by the Software Engineering Institute (SEI) of the Carnegie Mellon University, may be seen as the currently most established model. The SGMM aims at providing a management tool mainly for utilities. Developed in the USA originally by a consortium of utilities the model is structured along the integrated value chain of the power system. Accordingly the SGMM which as of August 2012 has reached version 1.2 proposes the following domains:

1. Strategy, Management, and Regulatory (SMR)
2. Organization and Structure (OS)
3. Grid Operations (GO)

4. Work and Asset Management (WAM)
5. Technology (TECH)
6. Customer (CUST)
7. Value Chain Integration (VCI)
8. Societal and Environmental (SE)

The maturity of an organization is assessed within each of the domains as they group Smart-Grid-related capabilities. The model mainly covers the network assets, processes and services, customer interfaces and customer interactions of the enterprise. Using these domains, the model aims at supporting mainly integrated utilities covering the generation, transmission and distribution of electrical energy. Within each domain of the model, maturity is separated into six levels from 0 to 5 as shown in Table 2.1.

Table 2.1 Maturity levels of the SGMM defined in [21]

Level 5 *Pioneering*	Organization is breaking new ground and advancing the state of the practice within a domain
Level 4 *Optimizing*	Organization's Smart Grid implementation within a given domain is being tuned and used to further increase organizational performance
Level 3 *Integrating*	Organization's Smart Grid deployment within a given domain is being integrated across the organization
Level 2 *Enabling*	Organization is implementing features within a domain that will enable it to achieve and sustain grid modernization
Level 1 *Initiating*	Organization is taking the first implementation steps within a domain
Level 0 *Default*	Default level for the model

The assessment using the SGMM is conducted by taking part in the SGMM navigation process ([22]). This process includes workshops and a survey in which the participants answer questions regarding the implementation of Smart Grid programs within the enterprise regarding the eight domains. This leads to the meta model of the SGMM shown in Figure 2.15. The *Organization* is assessed within the eight *Domains* the model consists of. The *Rating* within each domain is obtained by the answer towards the *Domain-Specific Questions*. Each of those questions addresses an *Expected Characteristic* of a domains Maturity *Level*. The maturity levels build on each other so a rating of level 3 within a domain implies the capabilities connected to the lower levels 2 and 1 are fulfilled by the enterprise. The maturity within a domain starts at the *Default Level*. This level is not connected with characteristics for the enterprise to fulfill and solely acts as the entry point of the model. The characteristics connected to a maturity level respectively depend on each other regarding their functional and organizational aspects. The maturity level descriptions within the survey are complemented by *Informative Characteristics* which are not part of the rating mechanism but help participants to understand the context of the level. While the rating and the domain-specific questions of the survey are directly connected with the underlying model definition of the SGMM, the survey further contains nonspecific questions which cover statistical data about the participants. This way the usage of the survey and the SGMM itself can be evaluated.

As the SGMM was mainly designed for integrated utilities in the USA, the model requires adaption to be applied in Germany or other countries with deregulated structures within the power system. In [15], an approach is introduced showing how maturity models like the SGMM may be configured.

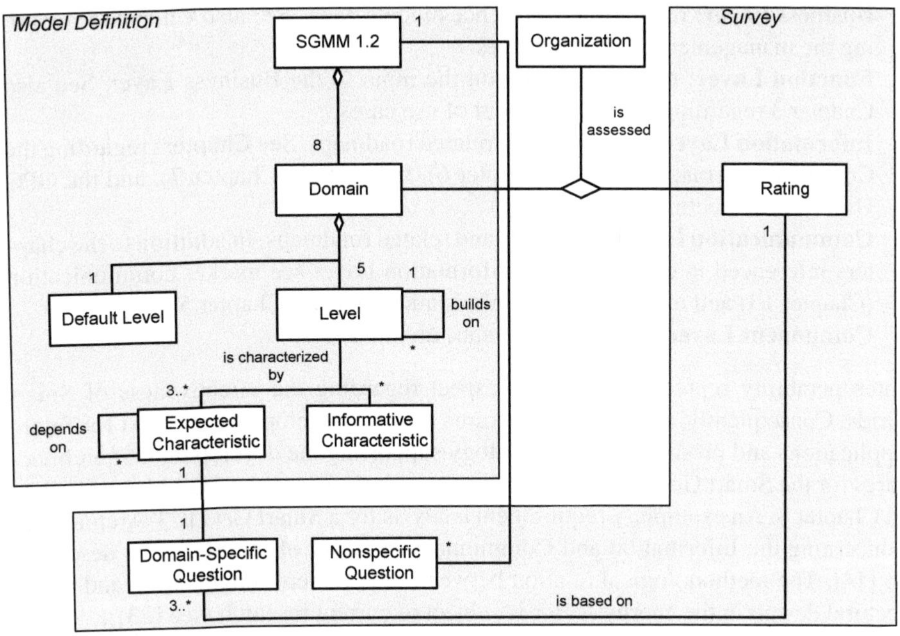

Fig. 2.15 Metamodel of the SGMM (adapted from [21])

2.7 Summary and Outlook

This chapter motivated the need for Requirements Engineering methodologies in the context of Smart Grid development (Section 2.1). It presented the basic principles and concepts of requirements engineering stemming from the Software Engineering domain (Section 2.2) before introducing the layered model of the GridWise Interoperability Framework (Section 2.3) as an approach to structure requirements analysis within the energy sector regarding interoperability and standardization aspects. Subsequently the SGAM was introduced (Section 2.4) which extends the structure of interoperability layers by additionally using domains and zones. The application of the SGAM matrix was illustrated by exemplarily outlining requirements regarding the technological field AMI at each layer of the model. With respect to the expected duration of the power systems migration towards Smart Grids, finally technological roadmaps and maturity models were introduced as means to manage long-term Smart Grid requirements and to assess the maturity of stakeholders (Section 2.6).

To derive Smart Grid requirements applying the methodologies outlined in this chapter, the use of additional information models and sources is recommended. The following sources may, among others, further support the SGAM layer oriented analysis of Smart Grid requirements:

- **Business Layer:** regulations, reference models, laws. See also Chapter 5 regarding the management of these sources.
- **Function Layer:** use cases based on the input to the Business Layer. See also Chapter 3 regarding the management of use cases.
- **Information Layer:** standards and related roadmaps. See Chapters regarding the Common Information Model (Chapter 6), IEC 61850 (Chapter 7), and the OPC Unified Architecture (Chapter 12).
- **Communication Layer:** standards and related roadmaps. In addition to the chapters referenced in context of the Information Layer see market communication (Chapter 13) and management of information models (Chapter 5).
- **Component Layer:** technical reference architectures.

Interoperability represents a critical aspect regarding the effectiveness of Smart Grids. Consequently, suitable architectures shall be developed as context for future applications and products. A methodology supporting the development of architectures for the Smart Grid using a layered approach similar to the SGAM is described in Chapter 4. An exemplary requirement analysis for a Smart Grid ICT-Architecture concerning the Information and Communication Layer of the SGAM is described in [14]. The methodological relation between requirements engineering and architectural design in the energy sector is subject to current research (see [23]).

References

1. Appelrath, H.J., Kagermann, H., Mayer, C. (eds.): Future Energy Grid - Migrationspfade ins Internet der Energie. Springer, Heidelberg (2012)
2. Boehm, B.: A spiral model of software development and enhancement. SIGSOFT Softw. Eng. Notes 11(4), 14–24 (1986)
3. Bruinenberg, J., Colton, L., Darmois, E., Dorn, J., Doyle, J., Elloumi, O., Englert, H., Forbes, R., Heiles, J., Hermans, P., Kuhnert, J., Rumph, F.J., Uslar, M., Wetterwald, P.: Smart Grid Coordination Group Technical Report Reference Architecture for the Smart Grid Version 1.0 (DRAFT) 2012-03-02. Tech. rep., CEN, CENELEC, ETSI (2012)
4. Bundesnetzagentur: "Smart Grid" und "Smart Market" Eckpunktepapier der Bundesnetzagentur zu den Aspekten des sich verändernden Energieversorgungssystems. Tech. rep. (2011)
5. Dänekas, C., Rohjans, S., Wissing, C., Appelrath, H.J.: "Future Energy Grid" - Migration Paths into the Internet of Energy. In: Cunningham, P., Cunningham, M. (eds.) eChallenges e-2011 Conference Proceedings. IIMC International Information Management Corporation Ltd., Florence (2011)

6. ENTSO-E and EDSO: The European Electricity Grid Initiative (EEGI) - Roadmap 2010-18 and Detailed Implementation Plan 2010-2012. European Network of Transmission System Operators for Electricity (ENTSO-E) (2010)
7. Hood, C., Wiedemann, S., Fichtinger, S., Pautz, U.: Requirements Management The Interface Between Requirements Development and All Other Systems Engineering Processes. Springer, Berlin (2008)
8. IEC: 62357 Second Edition. TC 57 Architecture - Part 1: Reference Architecture for TC 57 - Draft (2009)
9. IEC: 61968-100 (Draft): Application integration at electric utilities - System interfaces for distribution management - Part 100: Implementation Profiles for IEC 61968 (2011)
10. International Organization For Standardization: ISO/IEC 7498-1:1994 Information Technology - Open Systems Interconnection - Basic Reference Model: The Basic Model (1996)
11. Kotonya, G., Sommerville, I.: Requirements Engineering: Processes and Techniques, 1st edn. John Wiley & Sons (1998)
12. Kruchten, P.: The Rational Unified Process: An Introduction, 3rd edn. Addison-Wesley Longman Publishing Co., Boston (2003)
13. NIST: NIST Framework and Roadmap for Smart Grid Interoperability Standards (2010)
14. Rohjans, S., Dänekas, C., Uslar, M.: Requirements for Smart Grid ICT Architectures. In: 3rd IEEE PES Innovative Smart Grid Technologies (ISGT) Europe Conference (2012)
15. Rohjans, S., Uslar, M., Cleven, A., Winter, R., Wortmann, F.: Towards an Adaptive Maturity Model for Smart Grids. In: 17th International Power Systems Computation Conference (2011)
16. Royce, W.W.: Managing the development of large software systems: concepts and techniques. In: Proc. IEEE WESTCON, Los Angeles, pp. 1–9 (1970)
17. Schienmann, B.: Kontinuierliches Anforderungsmanagement Prozesse-Techniken-Werkzeuge. Addison-Wesley (2002)
18. The GridWise Architecture Council: GridWise Interoperability Context- Setting Framework. Tech. rep., The GridWise Architecture Council (2008)
19. The GridWise Architecture Council: GridWise Interoperability Context- Setting Framework. Tech. rep. (2008)
20. The Open Group: TOGAF Version 9 - The Open Group Architecture Framework (TOGAF), 9th edn. (2009)
21. The SGMM TEAM: Smart Grid Maturity Model - Model Definition Version 1.2 - A framework for smart grid transformation. Tech. rep., Software Engineering Instutite (2011)
22. The SGMM TEAM: Smart Grid Maturity Model - SGMM Compass Assessment Survey - A survey-based assessment of smart grid maturity. Tech. rep., Software Engineering Institute (2011)
23. Trefke, J., Dänekas, C., Rohjans, S., González, J.M.: Adaptive Architecture Development for Smart Grids Based on Integrated Building Blocks. In: 3rd IEEE PES Innovative Smart Grid Technologies (ISGT) Europe Conference (2012)

Chapter 3
IEC/PAS 62559-Based Use Case Management for Smart Grids

Jörn Trefke, José M. González, and Christian Dänekas

Abstract. Use cases are gaining momentum within requirements engineering for Smart Grids as they enable the description of how a system behaves in relation to its stakeholders. They allow to handle the complexity of systems and processes involved in Smart Grids. As more and more use cases for Smart Grids are developed, adequate management and coherent descriptions of use cases become necessary. The IEC Publicly Available Specification (PAS) 62559 aims at providing a method to develop use cases for Smart Grids and provides a corresponding template. Within this chapter, the need for use case management is motivated and proposals for a methodical enhancement as well as a tool support—both based on the IEC/-PAS 62559—are provided.

3.1 Introduction to Use Case Modeling for Smart Grids

Use Cases originated in object-oriented software engineering in the 90s and are widely used as a part of requirements engineering. Here, they serve to identify requirements considering a particular system under design, describing it from a technology-neutral viewpoint. A use case describes possible scenarios a system has to fulfill, i.e. they are a viewpoint on a system identifying related actors and functionality, e.g., for energy management systems (EMS) using Common Information Model (CIM)-based communication. They usually describe a particular goal an actor has and which is to be supported by the considered system resulting in a success or failure.

The presentation of use cases may be informal, using a text document supported by drawings, or more formal using Use Case Diagrams as specified in the Unified Modeling Language (UML). Descriptions of use cases can accordingly be created in

Jörn Trefke · José M. González · Christian Dänekas
OFFIS – Institute for Information Technology, Escherweg 2, 26121 Oldenburg, Germany
e-mail: {trefke,gonzalez,daenekas}@offis.de

M. Uslar et al.: *Standardization in Smart Grids*, POWSYS, pp. 39–57.
DOI: 10.1007/978-3-642-34916-4_3

different levels of detail/granularity, sometimes detailing even technical processes. They support the discussion of requirements between stakeholders of a system, so afterwards data models, interfaces, exchange processes or protocols may be chosen or derived. In addition to that, their need for standardization can be analyzed in order to create respective standards. This enables interoperability between actors and systems and shall create potential for innovative solutions for the Smart Grid. In the context of Smart Grids, the IEC Publicly Available Specification (PAS) 62559 [7] provides a template for use cases, which is currently being standardized, and also a methodology describing how to develop requirements using use cases.

This chapter describes an approach for use cases in the context of the Smart Grid and standardization, comprising general guidelines, a template, methodologies and tool support based on the IEC/PAS 62559. The contents of this chapter are based on work done for the German electrotechnical standardization organization DKE as well as further activities within national and international standardization and research at OFFIS. The approach has been discussed in the context of German national standardization and is also being considered in the M/490 mandate of the European Commission (EC).

Besides general use cases relevant to standardization, the contents of this chapter in parts also support development and management of enterprise-specific use cases. By linking the standardization and enterprise-perspectives, companies may analyze their use cases regarding relevant standards in order to provide interoperable technical solutions.

The remainder of this chapter is structured as follows: In Section 3.2 requirements for the construction and management of use cases are given. These provide the basis of the methodology presented in Section 3.3. Its structural concept, including conventions, template and repository structures as well as a classification system is introduced in Section 3.3.1. These structures are complemented by process models regarding the management of use cases in context of international standardization (see Section 3.3.2) and the construction of the use cases themselves (see Section 3.3.3). Before concluding with a summary and an outlook on future work, Section 3.3.4 presents the tool-support for the methodology.

3.2 Requirements for Smart Grid Use Case Descriptions

The Smart Grid is a complex system involving various actors and systems. Use cases may provide a common understanding regarding requirements and solutions ranging from descriptions of actors and functions to interfaces and data models. Actors addressed by use cases in the context of Smart Grids include among others:

- Companies in the energy industry
- IT manufacturers

- Equipment manufacturers
- Standardization organizations
- Legislators
- Companies from other sectors than the energy industry

The development of use cases, affecting this extensive group of stakeholders, requires participants with equally varying background knowledge and viewpoints. If use case descriptions shall express requirements and system functionality, and support collaboration between actors from different disciplines, like electrical engineers and IT experts, a shared methodology and tool support is required. This way standards and interoperability/conformance tests may also be based on use cases. National use case descriptions should be used to compile profiles to support the application of international standards in a national context conforming to regulatory provisions. In this context, the various levels of national and international standardization and the associated documents are to be observed.

The management of a large number of use case descriptions, originating from different specialized backgrounds, can be facilitated using a common structure of documentation to maintain consistency. Use cases must be of assured high quality and should be easily retrievable in order to create additional value in context of the initial documentation effort. As further use cases will be added and existing use cases be modified after their initial creation, their maintenance requires suitable classification criteria. These criteria should be robust and consistent over time in order to avoid the need for reoccurring classification. Consistency regarding the descriptions of use cases themselves must also be assured in this context, even if there are several editors. This can be achieved by using a template for use case descriptions and glossaries containing defined terms to be used. Regarding the different types of editors listed above, different types of templates may be required. The planned process of use case development and management should therefore be conceptualized accurately, as it impacts on the classification criteria, templates, and supporting tools.

As the development of use cases represents a collaborative task, it technically requires the support of multiple users, including roles and access control, locking, commenting, release, and configuration management of use cases. Since a large number of use cases is expected within the international standardization process, the classification, grouping, searching, and navigation of use cases should be supported. Further, uniform semantics should be used, redundancies be avoided, and descriptions be complete. Consistency may in this case be achieved by using a comprehensive model. Additionally multiple languages could increase acceptance within international standardization and projects. Ideally, the entire process of use case development should be supported by an integrated tool, which allows interlinking information across multiple use cases. With respect to the possibly considerable cost of such an approach, import and export interfaces to integrate and use other applications are an appropriate means to cope with that.

To avoid costs resulting from late changes to the use case process and template regarding its application in the international context, a broadly accepted template discussed in context of international standardization organizations should be used. The approach presented in this chapter is currently discussed within the "Sustainable Processes" working group in the context of the SG-CG M/490 mandate in Europe.[1] Appendix C includes an example using such template.

To conclude, the following measures have to be regarded, in order to structure the process for use case development and management appropriately:

- *Use of conventions:* Guidelines shall be introduced regarding naming conventions and the choice of specific (UML-) diagram types for the representation of real-world aspects.
- *Limitation of contents:* Avoidance of redundancies and reduction to relevant information only. This implies the consideration of related approaches, like the German Standardization Roadmap [4] for its classification of standards, the NIST conceptual model [9] or the German Information Technology Society's (within VDE[2]) views on domains related to the development of the energy information network [2].
- *Usage of familiar approaches:* E.g., selection of familiar and proven modeling languages, like the UML, which reduce the effort to "learn" use case development.
- *Tool support:* Reduction of manual work by software tool support, especially for quality and consistency assurance. Since the concepts, templates and classification criteria of the use case management solution will likely require extension or adjustment over time it should be possible to adopt these changes mostly in the background with little impact on the users and without invalidation of existing documents.
- *Exchange of Information:* Information shall be exchanged with other (standardization) organizations also concerned with the development of use cases for Smart Grids, like the FG Smart[3] group at International Telecommunication Union Standardization Sector(ITU-T).

3.3 Smart Grid Use Case Methodology

This section describes an approach for the development and management of Smart Grid use cases which aims at satisfying the requirements stated in the previous section. It is based on IEC/PAS 62559 [7] and provides extensions and a higher level of detail where needed. The description of the approach is divided into three parts. First measures for use case structuring and organization provided by the approach

---

1 See `http://www.cen.eu/cen/Sectors/Sectors/UtilitiesAndEnergy/SmartGrids/Pages/WGSustainable_Processes.aspx`

2 See `http://www.vde.com`

3 See `http://www.itu.int/en/ITU-T/focusgroups/smart/Pages/default.aspx`

are introduced, among others including an outline of the repository and template structures. This is followed by the description of the use case development and management process. It starts at the superordinate level including the application of use cases in standardization and is followed by an in detail description of the use case development process itself. The section concludes by outlining the support of these processes by a web-based software application.

3.3.1 Structuring and Organization Concept

The use case methodology provides the following measures to structure and organize use cases for Smart Grids:

- *Conventions and guidelines:* Conventions and guidelines ensure consistency throughout the development and management process.
- *Central repository:* A central location, which stores elements used during the creation and maintenance of use case descriptions. This includes, among others, definitions, actors, conventions, and relevant documents.
- *Template for use case development:* The internal template for use case description contains every attribute without restriction to a specific stakeholder role. It is advisable to restrict the application of the general template to use case experts and administrators. For regular users, a role-specific template should be provided instead. This reduces the complexity and ensures conceptual clarity within use case development.
- *Organization and classification of use cases:* Rules for organization and classification of use cases, which improve accessibility and application, especially for large numbers of use cases.

Conventions and Guidelines

In context of use case development, terms should be used in a consistent manner. Among others this especially applies to the following entities:

- Actors
- Roles
- Classification criteria
- Structuring elements
- Elements of the reference architecture

To ensure consistency, the use of established and accepted terms should generally be preferred to the invention of new terms. New terms should be submitted to a person in charge (e.g., a use case manager) who first should try to identify equivalent, existing terms. If needed, the adoption of a new term should be decided by consensus of several responsible persons to ensure acceptance and correct understanding of the term.

A use case gains comprehensibility and acceptance, if its description is, as far as possible, neutral regarding specific technologies, products, companies or projects. This should be especially regarded in context of terms used in the use case description. Terms established in the community should always be preferred to enterprise-, group- or project-specific terms. Regarding international acceptance the guidelines of the IEC should be followed.

Like the terms, also modeling languages and diagram types should be chosen in consensus to ensure a consistent description. The application of corresponding guidelines increases the understandability of descriptions. To accelerate the initial creation of use cases these rules and conventions might be mitigated. In such cases the use cases' compliance should be ensured within a subsequent revision. However, as new insights will be gained in the later application of the conventions and guidelines, they must be managed and extended regularly.

Central Repository

The repository will be used to centrally organize all relevant artifacts used in context of the proposed approach. Its structure is outlined in the following:

- Glossary
  - Terms, List of Roles (Market Roles / System Roles), Actors List, Acronyms
- Information Exchange
  - Data Objects, Data Protection Classes, Characteristics/Technical Requirements of Information Exchanges
- Structuring
  - Domains, Use Case Clusters, High-level Use Cases, Classification Criteria for Use Cases, Viewpoints for Use Cases
- Documents
  - Resources
- Methodology
  - Used Concepts/Conventions, Used Verbs
- Process
  - Used Roles Within Use Case Creation, Approval Status

Template

The template, as an agreed structure to document use cases, is based on IEC/-PAS 62559 [7] and was extended to a certain degree. The elements of the template are depicted in Figure 3.1.

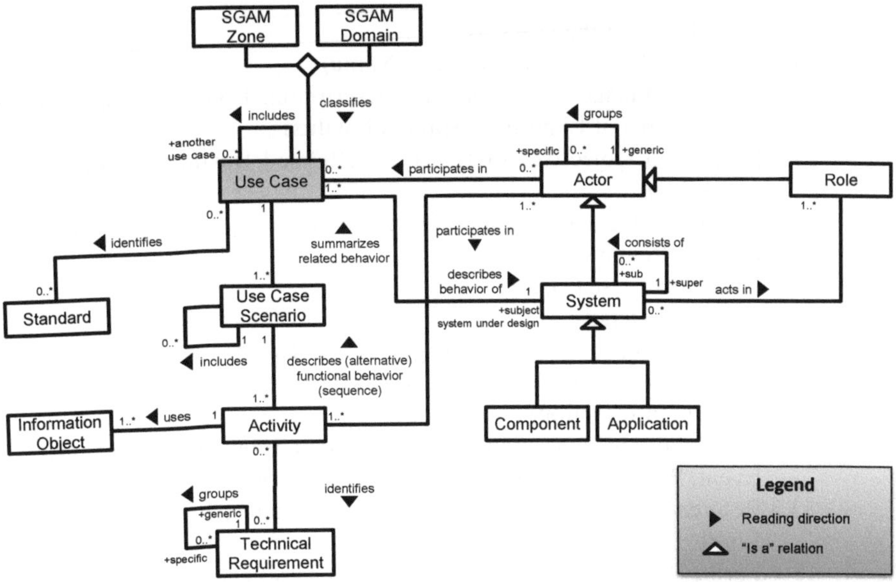

Fig. 3.1 Extended use case template meta data model following [7]

In this context *Use Cases* describe the expected goals an *Actor* wants to achieve with a particular *System* which is under design. This goal can involve other *Actors* required to achieve the goal, like for instance *Systems*, humans, *Applications* or *Components*. These act in a particular role depending on the use case. According to IEC/PAS 62559, use cases can consist of several *Scenarios* which can for instance describe a successful execution of the use case, or an error or maintenance scenario. Scenarios consist of several *Activities*, which define information exchange between actors including defined *Information Objects*. Each activity can moreover identify *Technical Requirements*, for example related to quality of service or security. With the extended version of the template, use cases can additionally specify standards to be applied for the realization of the use case. Use cases can also be classified to several criteria as for example the location in the Smart Grid Architectural Model SGAM (*Domains* and *Zones*—see Section 2.4 of the "Requirements Engineering for Smart Grids" chapter).

An outline of the use case template structure is shown below—an example according to this template can be found in Appendix C. The template consists of two parts. The *General* part contains attributes for rough description of a use case. It is complemented by the *Details* part which contains more specific information, focusing mostly on IT aspects.

General:

1 Description of the Use Case
1.1 Name of Use Case (ID, Domain, Name)
1.2 Version Management (Changes Description, Date, Author, Domain Expert, Area of Expertise, Approval Status)
1.3 Scope and Objectives of Use Case (Related Business Case, Scope, Objective)
1.4 Narrative of Use Case (Short Description, Complete Description)
1.5 General Remarks
2 Diagram of Use Case
3 Technical Details
3.1 Actors: People, Systems, Applications, Databases, Power System other Stakeholder (Name, Type, Description, Group, Further Use Case Specific Information)
3.2 Pre-conditions and Assumptions, Post-conditions and Events (Actor, Event, Pre-condition, Post-condition, Assumption)
3.3 References/Issues (No., Type, Reference, Status, Impact, Originator, Link)
3.4 Further Information to the Use Case for Classification/Mapping (Relation to Other Use Cases, Level of Depth, Prioritization, Applicability, Viewpoint, Keywords)
4 Step by Step Analysis of Use Case
No., Scenario Name, Primary Actor, Trigger, Pre-condition, Post-condition
4.1 Activities – Normal Scenario (No., Event, Name, Description, Service Type, Information Producer, Information Receiver, Information Exchanged, Requirement ID
4.2 Activities – Alternative, Error Management, and/or Maintenance/Backup Scenario (No., Event, Name, Description, Service Type, Information Producer, Information Receiver, Information Exchanged, Requirement ID
5 Information Exchanged (Name, Description, Requirement ID)
6 Common Terms and Definitions

Organization and Classification

The organization of use cases on the one hand supports navigation and thereby enhances accessibility for potential users. Classification criteria on the other hand are mainly intended to facilitate the identification of suitable use cases regarding a specific problem. The following tree structure can be used to organize use cases:

- *Use Case Cluster:* Represent a grouping of high-level use cases which may span several domains.
  - *High-level Use Case:* Abstract and generic use case which can be detailled by (multiple) other use cases.

- *Use case:* Description of a specific detailed use case consisting of activities.
    - *Activities:* Description of the detailed steps performed within a use case/use case scenario.

Regarding the structure presented above, a use case should be assigned only to one domain, while a use case cluster provides a cross-domain ordering characteristic. The implementation of the structure regarding instances of domains and use case clusters is dependent on the quantity of use cases contained in the repository. The same level of detail must be maintained on each level wherever possible. An adjustment of the hierarchy will probably become necessary over time.

3.3.2 Management Process

This section covers the management process for Smart Grid use cases (see also [8]), which integrates the structuring and organization concept outline in the previous section. The process is depicted in Figure 3.2 and ranges from the initial proposal for a project or use case to the specification of detailed use cases and interoperability tests. According to [8], the level of detail and realization increases along the stages *Ideas/Requirements*, *Elaboration* to *Standards Development*.

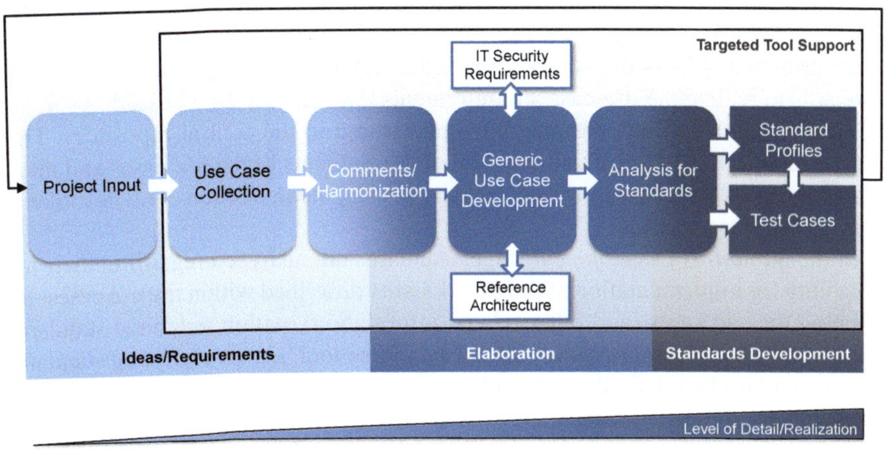

Fig. 3.2 Management process for Smart Grid use cases based on [8], [11]

Within the *Ideas/Requirements* stage, project proposals for standardization are elaborated within technical committees or working groups. This work usually comprises short descriptions of the project and corresponding use cases on a high level. The developed use cases are stored within the central repository and are documented based on the IEC/PAS 62559-based template. Use cases are developed in a two-step approach, in which initial functional descriptions are provided by domain experts and enhanced with more details by technical experts. Also, existing use cases from different sources should be considered. To integrate theses use cases, they should be consolidated and harmonized by standardization experts in terms of used concepts and the application of the template beforehand. Further on, use cases are classified and structured to support retrieval and analysis in later stages. The classification may be conducted according to the level of detail (e.g., high-level or detailed), the applicability within a geographical context or its maturity (see recommendations in [1]).

The focus in the *Elaboration* stage is on a more detailed and precise description of the generic use cases as well as the identification and definition of interdependencies to other artifacts (e.g., other use cases or actors). The generic use cases developed in this stage are defined on the basis of similarities between related use cases which have already been analyzed. Generic use cases should abstract from specific conditions as far as possible (e.g., from country-specific regulations or laws) in order to be used in the standardization context. Artifacts defined in a generic use case, like information objects and actors (e.g., systems, software applications or other components), can provide valuable input for a reference architecture. Such a reference architecture can guide the implementation and also serve as a reference ontology for further use case creation. Further, the elaboration of the classification, e.g., regarding the interoperability layers as defined in the Smart Grid Architecture Model (SGAM) [3] (see also Chapter 2 for details on the SGAM), can take place at this point. In addition to that, the security requirements can be defined at this stage. This can take place on the basis of a use case scenario analysis with the assistance of the information objects and actors identified. With the information identified so far, standards to realize the use case's requirements may already be assessed. As in the previous stage, all results are integrated and stored in the central repository. This enables an interlinked information model of the elements of the use cases and interrelated artifacts. Regarding the step *use case development*, Section 3.3.3 describes a proposed process in detail.

In the *Standards Development* stage, standards are analyzed regarding their applicability for implementation of the functionality described within the use cases. By mapping the use cases (and other related information), within a defined standards framework (as, e.g., planned with the IEC mapping tool[4]), gaps regarding standardization can be identified. This not only enables a gap analysis but also to analyze the coverage of existing standards by use cases. These integrated information also support the management of standards portfolios, i.e. depending on the identified gaps/coverage, standard revisions can be conducted or the discontinuation of standards can be decided. Standards which are used or were at least identified to be used in conjunction within in a particular use case, can be grouped in subsets (standard profiles). These profiles shall ease the standard-based implementation of this use case. As the use cases shall serve for the definition of standards in the proposed ap-

[4] See `http://www.iec.ch/smartgrid/`.

proach, they can also be used as input for the definition of interoperability tests for standards or standard profiles later on.

With an advancing vision and implementation of Smart Grid functionalities, new requirements will arise. Consequently, this results in the definition of new use cases and also in the change of existing ones. Beyond that, this means that standards developed according to these use cases will have to be revised and thus, the whole process will have to be executed again with these changed inputs (line).

Not least these previously mentioned effects, lead to the conclusion that the effort to run such a process increases with the number of use cases and standards to be examined respectively. Executing this process manually, especially checking conventions, consistency and cascading changes, will probably introduce errors and be inefficient. Here, information technology (IT) can be used to facilitate the use case creation, administration, analysis, and exploitation process, e.g., by assuring consistency and the compliance to conventions.

A tool support which targets these steps is proposed later in Section 3.3.4 and highlighted in Figure 3.2 (box "Targeted Tool Support"). Based on the general methodology outlined in this section the next section focuses on a proposal for a detailed use case development.

3.3.3 Use Case Development and Application Process

This section outlines the detailed development and application process regarding Smart Grid use cases. As depicted in Figure 3.3, the process distinguishes three roles, while dividing the activities connected to them into five phases. These roles and phases are further elaborated in the following.

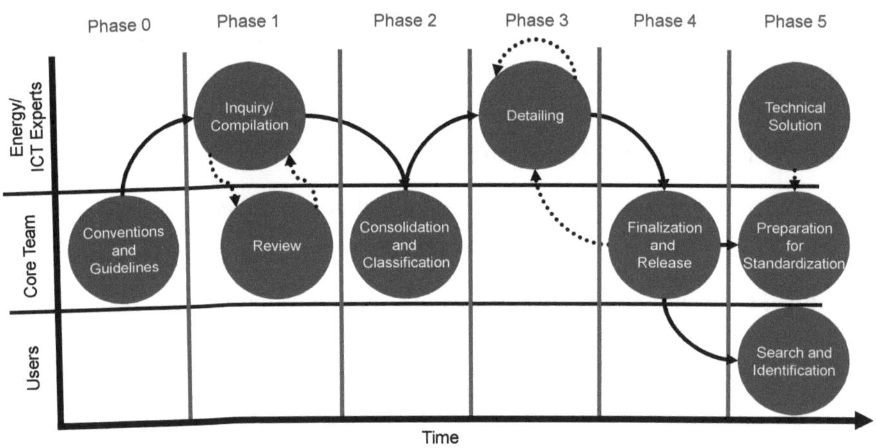

Fig. 3.3 Process for use case development and processing

Actor Roles

In short *Energy and ICT experts* essentially cover the development of the use cases and apply their practical experience. More specifically *Energy experts* possess far-reaching expertise in a particular discipline of power system engineering and contribute substantive domain-knowledge to the use case. They identify new use cases and initiate their development. *IT experts* in contrast possess technical expertise regarding modeling and implementation of information and communication technologies. Their domain-knowledge regarding the power system is mostly general. They therefore shall provide technical details for the use case and estimate the feasibility regarding an appropriate ICT implementation.

The *Core Team* in turn conducts the management of the use cases. This includes the specification of conventions and guidelines, the review of use cases regarding their individual quality and overall consistency, their finalization and release of the use cases as well as the preparation for standardization. The team consists of interdisciplinary experts, whose individual domain-specific knowledge enables them to assure the quality and relevance of the use cases. The broad expertise on the team level further enables it to classify use cases in a holistic context. This context may be a professional domain model or a reference architecture. The core team is lead by at least one person responsible for the management of the use cases and communication of the teams decisions. Additionally, the core team includes standardization experts. They possess knowledge of established standards and current standardization projects in various standardization organizations, and can therefore assess and classify the relevance of new approaches in this field. Furthermore, they are familiar with the conventions and processes for submission of proposals to standardization organizations. Use case administrators assist the core team regarding the technical maintenance of the use case collection. If needed the team may seek additional support, e.g., by methods experts.

The *Users*, constituting the third and final role, finally shall be enabled to search for appropriate use cases, e.g., on the basis of their classification.

These roles will be used in the following to highlight the responsible actors within the five phases of the process model.

Phase 0: Conventions and Guidelines

Objective: In this phase, conventions and guidelines to govern use case development are established.

Procedure: Conventions and guidelines for use case development and management are elaborated by the core team. Among others, these affect the structure of the template used for documentation of the use cases, the graphical notation to be used, the structure of the central repository, and the criteria for use case classification.

Actors involved: Core team

Information required: Related approaches, target groups

Results: Conventions and guidelines for the use cases

Phase 1: Compilation

Objective: In this phase a first version of a use case is created. It shall comprehensibly describe the use intended for a system and provide information regarding the stakeholders involved on a technical level. It shall be ensured that the use case is relevant in context of the other existing use cases while keeping the overlaps with them at a minimum.

Procedure: The initial use case is normally compiled or contributed by stakeholders interested in standardization of its contents. The description of the use case shall follow IEC/PAS 62559 and the specific additions based upon it (see Section 3.3.1). This first version of a use case shall express stakeholder needs connected to it. The energy expert responsible for its description shall give a specific but easily comprehensible example, ideally motivated by a practical problem. The resulting use case outline is proposed to the core team which conducts a review. This review determines the acceptance and further elaboration of the use case. Review criteria include the thematic relevance of the use case in the context of the objective and its overlap with other use cases. The review process might be supported by a software tool. In case of thematic irrelevance or strong overlap with other use cases, the use case managers may reject the use case or require the authors to narrow down its topic. In case of acceptance by the core team, the use case is further elaborated on a technical basis. This includes the identification of the actors involved in the use case and their unequivocal definition if they are not already part of the global actor list. Furthermore, legal restrictions and requirements as well as precondition and assumptions shall be defined for the use case.

Actors involved: Energy experts, core team

Information required: Specific use case with need for technical support, list of defined actors, central repository

Results: Technical part of the use case (aims of the system, description of the use case, actors involved, restrictions and requirements, preconditions and assumptions)

Phase 2: Consolidation and Classification

Objective: In this phase, the terminology of a use case and its compliance with the conventions and guidelines are reviewed. Amendments or additions are made if necessary. Finally, the use case is classified within the overall context, enabling efficient evaluation and identification.

Procedure: The terminology, conventions etc., used in the technical part of the use case, are reviewed and if necessary adjusted to comply with the global conventions and guidelines. Terms of the use case not yet included in the central repository may be added if they represent essential new concepts. The use case is then enhanced with diagrams as required (using diagram types as stipulated, e.g., particular UML diagrams [5]). Existing, non-conforming diagrams are converted into a guideline compliant form. This shall ensure the consistency of graphical notations throughout the collection of use cases and contribute to ease of understanding for the readers. Furthermore, references to laws or regulations and to relevant standards and corresponding working groups shall be added at this stage. The core team afterwards classifies the use case on the basis of specified classification criteria (see Section 3.3.1).

Actors involved: Core team

Information required: Initial version of the use case from energy experts, central repository, list of actors

Results: Revised version of the technical part of the use case (adjusted terminology, references, classification)

Phase 3: Detailing

Objective: Addition of further details and workflows to the use case.

Procedure: The use case is provided with further details in this phase. This is done by IT experts who refine the use case by adding and describing activities. These activities represent the individual steps of the use case, which have to be performed in a defined sequence (i.e. workflow) to achieve the intended aim.
Alternative sequences of activities which may, for example, occur in the event of a fault, are also identified in this phase. For each individual activity, conditions and assumptions essential for the performance shall be defined. The identification of the individual activities and alternative sequences may result in the introduction of additional activities. The detailing phase may therefore have to be repeated, leading to an incremental refinement of the use case.

Actors involved: IT experts

Information required: Accepted, revised version of the use case from the use case managers, central repository, list of actors

Results: Detailed version of the use case (description of the individual activities in the use case)

Phase 4: Finalization and Release

Objective: This phase comprises the review of the use case and decision on further processing or release. The use case may go forward into standardization, prepared for standardization or be released to the interested professional public.

Procedure: The detailed use case is reviewed by the core team in this phase regarding completeness and also for compliance with conventions and guidelines. The use cases' terminology, conventions etc. are checked and if necessary adjusted for conformity. Terms not yet available in the central repository representing essential new concepts may be included in the central repository. Likewise, the use case is again enhanced with diagrams of stipulated type as required and existing diagrams are analyzed and adjusted for compliance with the guidelines. References to laws or regulations and to standards and corresponding working groups are reviewed and added, if required. Finally, the use case is released by the use case managers for the further standardization process or the preparation process for standardization. If needed, the use case is sent back for further processing (phase 3), e.g., to arrange for changes by the IT experts.

Actors involved: Core team (especially use case manager)

Information required: Detailed version of the use case from the IT experts, central repository, list of actors

Results: Final version of the use case suitable for the further standardization processes

Phase 5: Preparation for Standardization, Use and Identification

Objective: In this phase, the use case is extended for the relevant standardization organization and its subsequent use for implementation. Additionally, use cases in this phase may be located by users of the repository.

Procedure: From the finalized use cases and existing solutions developed by the energy and IT experts, the core team identifies parts to be standardized. These parts may serve as the basis for conceptual designs (e.g., for data modeling, interfaces, etc.). Alternatively, these parts may also be proposed by the energy and IT experts. The preparation of these parts for a specific standardization organization shall be conducted by a standardization expert. It satisfies the requirements of the standardization organization and goes forward into the further standardization process. Users can locate relevant use cases based on the classification within the repository and use or implement the use cases and/or the resulting standards. Considering a continuous use of standards up to implementation, dedicated test use cases and standard profiles may be developed subsequently. These shall be appropriately detailed to permit technical implementation and testing for interoperability where necessary.

Actors involved: Core team (especially standardization expert), energy and IT experts, repository users

Information required: Final version of the use case from the core team, standardization organization, and its requirements regarding the use cases

Results: Version of the use case prepared for the relevant standardization organization

3.3.4 Tool Support

In order to support the use case development process illustrated in the previous two sections, the use of a tool is recommended. Figure 3.4 outlines the architecture of the Use-Case-Management-Repository (UCMR) (see also [11]), which shall succeed to this task. The architecture basically consists of layers regarding presentation, functions, and data, according to the proven three-tier architecture pattern. Since it is suggested that stakeholders should be supported by individual views, the presentation layer includes stakeholder-oriented access to the use cases and the use case management functionalities.

The function layer consists of three functional groups, which enable the *initial creation*, *management*, and *publication* of use cases. For the collection and creation of use cases, a *Creation* functionality is provided. The content is—depending on the stakeholder group—specifically presented: domain experts will only see the basic information of the use case template they are required to enter, ICT experts for instance can access the presentation of more detailed aspects like use case scenarios and specify activities. Furthermore, *Import* functionalities enable to integrate external use cases to make them available in the UCMR. The *Administration* functionality can be used to classify and structure defined/collected use cases to organize them and enable a structured retrieval. Moreover, the process from initial creation to release of a use case should be managed. The quality aspect of use cases (e.g., level of completeness, use of glossary terms, links to other artifacts, compliance regarding guidelines and conventions) is addressed with the *Analysis/Quality Assurance* functionality. Providing analyses, this shall enable the development of consistent, high quality use cases.

Finally, functionality to *Export* use cases for the use in other tools is provided. On the one hand this can comprise UML, which best reflects the integrated, machine-processable information. Using this export format also allows its application within further model-based development efforts and eases the use case customization efforts by implementers using their own tool chain. On the other hand the export of text documents is possible, which are easily shareable and printable.

All information processed by the respective functionality is stored in the data layer, which basically is a repository. The repository is structured according to the use case template which was presented in Section 3.3.1 and Figure 3.1 in particular. This incorporates several enhancements in comparison to the IEC/PAS 62559 template, e.g., the classification of use cases to support their identification, as suggested in Section 3.3.1. In addition to that, data concerning related standards can be stored in order to support the gap analysis of standards, which is conducted in the last phase of the overall methodology.

By defining the explicit formalization of the template as the data model for the tool, all related information can be stored within an integrated data base. This especially means that use case descriptions with consistent, interlinked entities can be created. Defined entities like actors, use cases, information objects, particular scenarios' activities or technical requirements can be reused for the description of multiple use cases.

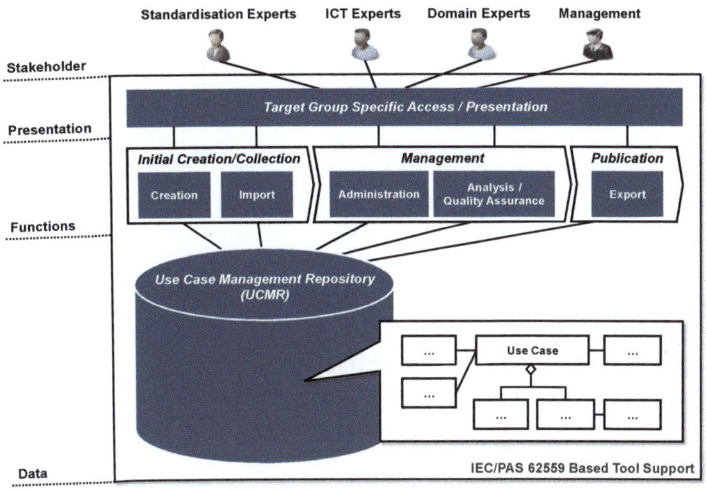

Fig. 3.4 Architecture outline of a tool to support Smart Grid use case management based on [11]

On the basis of the architecture outline depicted in Figure 3.4, a first version of an integrated tool to support use case management has been implemented. The tool (named Chronos Use Case Editor), was built on the basis of the open-source tool Chronos Web Modeller[5] (CWM) as a web-based tool, and developed by OFFIS in a joint project with IBM Germany and DKE. The Chronos Use Case Editor is used for modeling Smart Grid use cases within the EU Smart Grid Coordination Group's (SG-CG) working group Sustainable Processes [1]. This first version has proved feasibility, at least for a reduced set of features which have been implemented. Figure 3.5 shows a screenshot of the use case editor prototype.

The editor provides an area containing workspaces on the left side of the window, where the artifacts from the UCMR are stored in packages according to defined structures (e.g. working groups or individual user workspaces). These structures can

[5] See `http://sourceforge.net/apps/wordpress/olympos/`

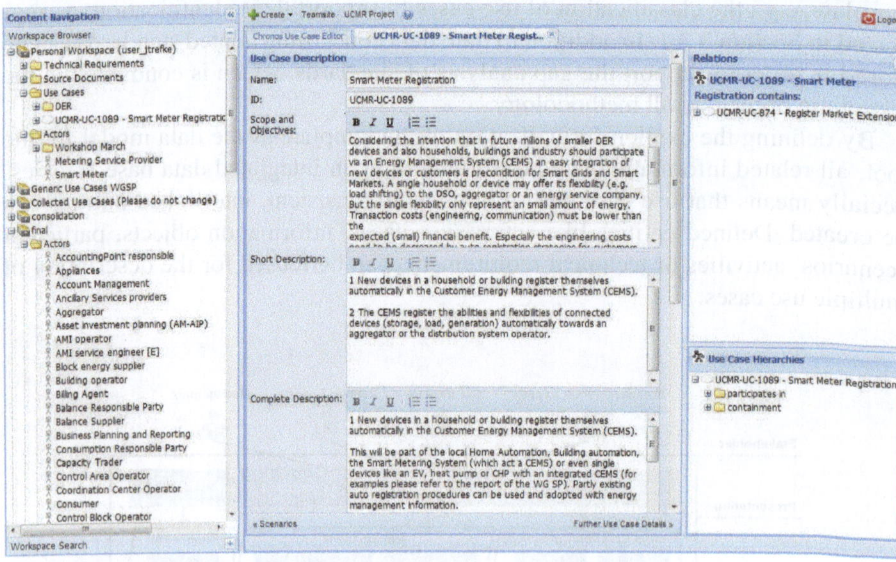

Fig. 3.5 Screenshot of the Chronos Use Case Editor prototype

be navigated and accessed as needed by the users to identify their required information (e.g., use cases or actors). The editing takes place in the central panel which displays input fields depending on the artifact opened, i.e. basically the information which is required to fill out the extended IEC/PAS 62559 use case template. With the panel on the right, links to use cases and other artifacts are displayed and can be created.

Users can be supported regarding the creation and editing of use cases or other artifacts by utilizing further information integrated in the repository. For example, they can drag existing actors from the workspace to applicable areas in the use case or select predefined information from lists, e.g. for classification purposes. Finally, use cases can be exported as text documents, but also as UML using a template-specific profile.

The presented approach so far proved helpful to support domain experts—regardless of their UML/modeling background—in defining use cases collaboratively (referring to the *Initial Creation* functionality from Figure 3.4). The first Chronos Use Case Editor prototype primarily focused on creation functionalities. Further development efforts considering import, management, and publication are subject of future work.

3.4 Summary and Outlook on Future Work

Within this chapter we motivated the need for a structured and tool-supported use case management within standardization and beyond. First requirements regarding case management in this context where outlined. To address these requirements, a methodology supporting the management of Smart Grid use cases was presented. This encompassed structural concepts regarding the development and organization of use cases as well as their integration into appropriate process models. In order to support this methodology, a tool support concept and prototype was outlined at the end of this chapter. A use case collection created using the methodology and an integrated tool, may be used to analyze the various interdependencies between Smart Grid functionalities and actors. With this contribution, standardization can therefore be supported regarding the definition of interfaces and respective standards.

Future work within OFFIS will focus on the enhancement of the existing tools and classifications to improve user support regarding the identification, creation, and analysis of use cases. In addition, further integration with existing approaches related to ICT-architecture development (see Chapter 4) or requirements engineering (see Chapters 2 and 5) are ongoing (see [6] or [10]) and envisaged.

References

1. Apel, R., Hourdouillie, R., Lambert, E., Postma, A., Sand, K., Stein, J., Strabbing, W., Tornelli, C., Trefke, J., Tusch, J., von Jagwitz, A.: Smart Grid Coordination Group Technical Report – Use Case Collection, Management, Repository, Analysis and Harmonization (Draft). Tech. rep., CEN CENELEC, ETSI (2012)
2. Benze, J., Blechinger, T., Diedrich, C., Hänchen, H., Honecker, H., Hübner, C., Khattabi, M.: Kieß ling, A., Krings, H., Lehnert, R., Lehnhoff, S., Rohjans, S., Stein, E., Tretter, R., Uslar, M.: ITG-Positionspapier Energieinformationsnetze und -systeme Teil A – Verteilnetzautomatisierung. Tech. rep., VDE ITG (2012)
3. Bruinenberg, J., Colton, L., Darmois, E., Dorn, J., Doyle, J., Elloumi, O., Englert, H., Forbes, R., Heiles, J., Hermans, P., Kuhnert, J., Rumph, F.J., Uslar, M., Wetterwald, P.: Smart Grid Coordination Group Technical Report Reference Architecture for the Smart Grid Version 1.0 (DRAFT) 2012-03-02. Tech. rep., CEN, CENELEC, ETSI (2012)
4. DKE: The German Standardization Roadmap E-Energy/Smart Grid. VDE (2010)
5. EPRI: Integration of Advanced Automation and Enterprise Information System Infrastructures - Harmonization of IEC 61850 and IEC 61970/61968 Models. Tech. Rep. 3, EPRI (2007)
6. González, J.M., Dänekas, C., Trefke, J., Uslar, M.: Supporting Interoperability in Smart Grids. In: I-ESA 2012 - Interoperability for Enterprise Systems and Applications (6th International Conference) (2012)
7. International Electrotechnical Commission (IEC): Publicly Available Specification (PAS) 62559 IntelliGrid Methodology for Developing Requirements for Energy Systems. Tech. Rep. 1.0 (2008)
8. Kellendonk, P., Kießling, A., Uslar, M.: Definition von Use Cases in der Normung - Basis für eine aktive Beteiligung privater Haushalte im Smart Grid on the Definition of Use Cases in Standardization. In: ETG Kongress 2011 (2011)
9. NIST: NIST Framework and Roadmap for Smart Grid Interoperability Standards (2010)
10. Trefke, J., Dänekas, C., Rohjans, S., González, J.M.: Adaptive Architecture Development for Smart Grids Based on Integrated Building Blocks. In: 3rd IEEE PES Innovative Smart Grid Technologies (ISGT) Europe Conference (2012)
11. Trefke, J., González, J.M., Uslar, M.: Smart Grid Standardisation Management with Use Cases. In: 2nd IEEE ENERGYCON Conference & Exhibition, Florence, Italy, pp. 966–971 (2012)

1. Appel, R., Hounbouh, R., Gendre, E., Postel, A., Stief, F., Weise, T., [illegible]

[illegible]

Chapter 4
Development of Smart Grid Architectures

Jörn Trefke and Christian Dänekas

Abstract. Because of new producer-, storage- and demand-side management systems which are introduced for a Smart Grid, new data pools, interfaces and processes, arise. Existing legacy systems have to interact with new systems, therefore, the functional and process logic of the power system will be distributed in a more complex way. Information and communication technologies will be required to realize these complex interactions. Developing such a complex system architecture requires a structured approach, which considers the various stakeholders' concerns. Accordingly, a fundamental architecture management of the system landscape as well as a process overview needs to be established by energy suppliers. This chapter presents an introduction and basics on this topic.

4.1 Motivation for Architecture Development

The development of large-scale systems—like the Smart Grid—is a complex task. It involves numerous stakeholders along the value chain, as for instance producers, utilities or consumers, as well as manufacturers, ICT-experts or even stakeholders from the automotive sector regarding electrical vehicles. These systems usually consist of various interrelated elements themselves and also relate to other, equally complex and heterogeneous systems. Therefore, they are hard to understand as a whole for individuals. In addition, the time from development to realization and operation spans over a long period of time and involves great costs. These facts lead to the assumption that a well-planed development of these systems is advisable. Each system's architecture has to be managed to support the development of a complex system like the Smart Grid.

Jörn Trefke · Christian Dänekas
OFFIS – Institute for Information Technology, Escherweg 2, 26121 Oldenburg, Germany
e-mail: {trefke,daenekas}@offis.de

M. Uslar et al.: *Standardization in Smart Grids*, POWSYS, pp. 59–77.
DOI: 10.1007/978-3-642-34916-4_4

Since the term *system* can relate to various subjects, as for instance devices, software or enterprises, and systems can be composed of subordinate systems, the meaning of the term architecture is dependent on the current context. In this sense, "Smart Grid architecture" comprises a wide range of architectures regarding the systems involved in the realization of a smart grid—a term which is not even clearly defined and whose subjects are often dependent from company-specific or regional goals. Therefore, existing recommendations for Smart Grid architectures differ regarding the level of detail, regional focus and organizational scope. Examples are the NIST conceptual models defined in [10], the Smart Grid Architecture Model (SGAM) defined in [3] or the Smart Grid Standards Architecture as defined in IEC 62357 [5].

Enterprises representing actors in a Smart Grid producing, trading, distributing and selling electricity, are now encouraged to put the Smart Grid into practice. However, the artifacts of the architecture models named above are not directly applicable for enterprises, i.e. they must be adapted to enterprise-specific requirements and require to be applied thoughtfully using a methodology which can leverage the already defined structures. Within the enterprise-specific context, several factors, depicted in Figure 4.1, lead to changes in different architecture views. Changes in the business context, as for instance changed business processes due to enterprise-external requirements (e.g. automated meter reading vs. manual meter reading) also affect other parts, like (software) applications and underlying technology. In this case, software applications must offer functionality supporting this process. Moreover, when reading meter data every 15 minutes, corresponding solutions able to process the large amount of data must be available. However, data will not be available unless the underlying technology provides them, i.e. in terms of meters. This requires the availability of reliable, digital solutions. Not only do innovations in business influence technology, but also new technology or applications can influence the business. Managing this process requires a holistic development approach, from requirements engineering to developing architectural building blocks and finally selecting (where applicable) and implementing solutions.

This chapter addresses the foundations of architecture and its description in Section 4.2, and presents these aspects in the enterprise context in Section 4.3. These architecture basics lay the foundation for a holistic architecture development method for enterprise architecture and its application in the context of Smart Grids, which is outlined in Section 4.4. Finally, this chapter ends with a conclusion and an outlook on the extended context in Section 4.5.

4.2 On Architecture

There are various definitions of system- and software architecture, and in practice the term is used manifoldly. Empirical research [12] identified at least four different metaphors associated with architecture. These are "architecture as a blueprint", "architecture as literature", "architecture as language" and "architecture as decision". In the first case this means according to [12], that architecture is a working implementation, where its description contains high-level concepts and serves as a plan for the structure to be implemented. The second meaning implies, that architecture is the solution or the collection of solutions made in the past and its description is seen as documentation oriented towards future readers to serve as reference and contains collected solutions. The third metaphor focuses on architecture as a common understanding, where its description serves as a common basis for communication among stakeholder groups and for achieving common high-level structures about the system. Finally, the fourth metaphor understands architecture as the basis for rational decision-making and its description captures the decisions about the structure of the system.

As can be seen, depending on the meaning of architecture—which differs among stakeholders—the contents of its description and its level of detail varies. The ISO/IEC/IEEE standard 42010 "Systems- and software engineering — Architecture description" [8] captures and integrates concepts around architecture and its description, and provides several terms used in this context, which are used in this chapter. These terms are valuable for discussing and creating architecture (descriptions), and facilitate the understanding of existing work. The architecture of a system according to [8] is described as "fundamental concepts of a system in its environment embodied in its elements, relationships, and in the principles of its design and evolution". Every system has an architecture, which may not always be made explicit or documented. Figure 4.2 shows a conceptual model of an architecture description's contents, which is presented in the following.

A system is built to achieve one or more specific purposes, which are realized by several parts (even other systems) that are interacting with each other. Systems can be of different natures, as for instance hardware, software-products or enterprises, and may be used in various domains. Everything outside a system is considered as its environment and can in particular influence the system and vice versa. In other words, this means that a system's boundary is defined by its environment.

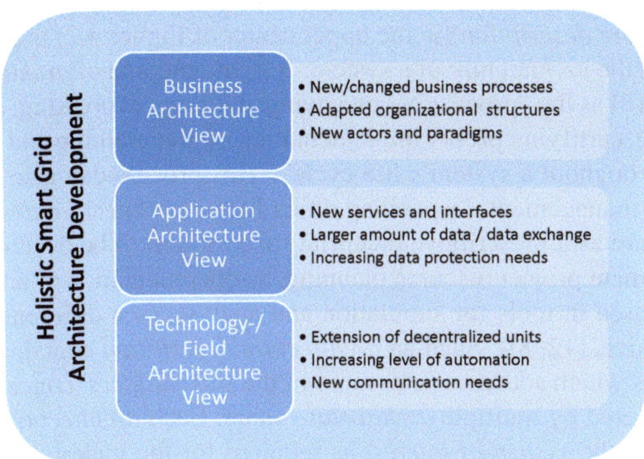

Fig. 4.1 Exemplary issues in Smart Grid architecture management on different layers

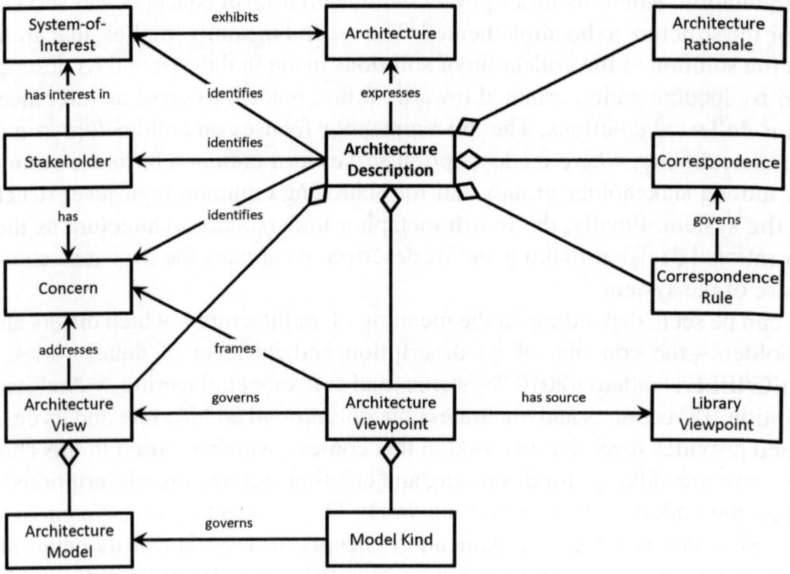

Fig. 4.2 Conceptual model for architecture description following [8]

The *system-of-interest* exhibits an *architecture*, which must not be necessarily documented. *Stakeholders* of a system have interests in this particular system. These could for example be contractees, developers, users or maintainers, who have different, in some cases even opposing architecture-related *concerns* and usually expect the *system-of-interest* to fulfill specific purposes. The main interest of a user will be, for instance, the functionality of the system to achieve a specific task, while developers are more interested in components and technical implementation details of the system.

An *architecture description* (in the upper center of Figure 4.2) is a work product that documents the *architecture* of a system. It is an outcome of *architecting*, which is described in [8] as the "process of conceiving, defining, expressing, documenting, communicating, certifying proper implementation of, maintaining and improving an architecture throughout a system's life cycle". This process can also be subsumed as architecture management. Depending on its focus, *architecture descriptions* can for instance serve as a prescriptive blueprint for a system to be developed, as a basis for development project resource planning, as documentation of an already built system, or be used in tools for simulation and analysis (see different meanings of architecture above, [12, 8]). Such an *architecture description* contains multiple *architecture views* which address one or more of the *stakeholders' concerns*; *concerns* can also be covered by multiple *architecture views*. Each *architecture view* depicts relevant parts of the *system-of-interest* as required for the underlying *concerns*. A "complete" view of the system's architecture will then be the consolidation of all architecture views which address the relevant stakeholder concerns. However, architecture descriptions in general are to present the key concepts which are relevant for the stakeholders at a specific point in time and so do not cover the complete complexity of systems and their architectures, but rather reduce it to relevant aspects.

An *architecture view* is governed by an *architecture viewpoint* which frames particular *concerns* (and identifies *stakeholders* for which these are relevant). Moreover an *architecture viewpoint* defines how to model in order create *architecture views*. This can, according to [8], include "languages, notations, *model kinds*, design rules and/or modeling methods, analysis techniques and other operations on views". *Model kinds* govern the *architecture models* which are used in (different) *architecture views*.

Architecture viewpoints can be defined to document a specific *architecture*, but they can also be defined outside the context of a specific *architecture description*. In the latter case, they are applicable in many architecture descriptions and are referred to as *library viewpoints*. Choosing *architecture viewpoints* for an architecture description—and creating respective *architecture views*—basically depends on the concerns/stakeholders.

Beyond the already identified elements, an *architecture rationale* is part of the *architecture description*. Here, the architecture rationale covers multiple aspects, like a rationale for each architecture viewpoint (e.g., its stakeholders, concerns and model kinds), a rationale for key architecture decisions and a rationale for choices made considering alternatives. In addition, an *architecture description* defines *correspondences* between elements used to construct the *architecture description* (AD elements). Identifying correspondences between AD elements documents consistencies and inconsistencies across multiple views and models. *Correspondences* should be governed by *correspondence rules* that allow the identification and analysis of consistencies and inconsistencies. Further recommendations for contents of architecture descriptions can be found directly in [8].

When repeatedly architecting systems of the same nature (e.g. software) within the same domain, it seems evident to reuse existing work products and best practices regarding the process as well as established viewpoints for architecture description. *Architecture frameworks* address this topic. Following [8], frameworks provide "conventions, principles and practices for the description of architectures established within a specific domain of application and/or community of stakeholders". *Architecture frameworks* are, e.g., useful for creating architecture descriptions, communicating about architecture, and implementing tools to support development. In general, they provide an aligned set of viewpoints, with respective information (stakeholders, concerns, model kinds, etc.). Examples of frameworks include the 4+1 View Model [9], The Open Group Architecture Framework (TOGAF) [13] or the Reference Model for Open Distributed Processing (RM-ODP) [7]. This can simplify the design and development of multi-stakeholder architectures.

The contents of architecture *frameworks* are shown in Figure 4.3. An *architecture framework* identifies several *stakeholders* which are relevant for the particular application domain as well as their *concerns*. In order to give recommendations on how to describe architectures in this domain, these *concerns* are framed by

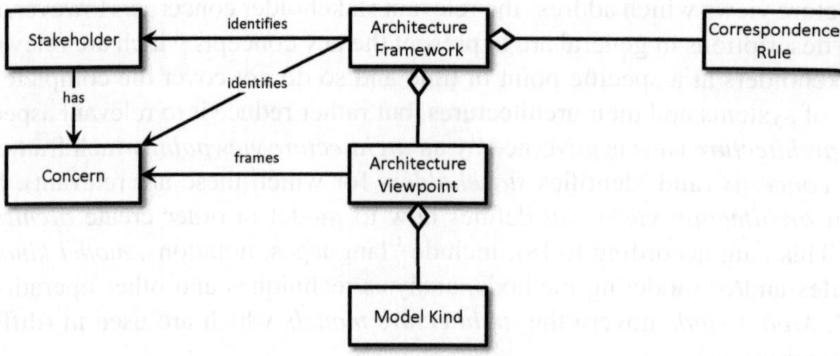

Fig. 4.3 Contents of an architecture framework following [8]

architecture viewpoints. Analogous to the previous explanation, *model kinds* are defined for *architecture viewpoints* in order to create architecture models. *Correspondence rules* define restrictions on relations between AD elements (e.g. *stakeholders*, *concerns*, *viewpoints*) and allow to analyze correspondences. Additionally, the ISO/IEC/IEEE 42010 standard [8] defines further criteria for *architecture frameworks*, their adherence to *architecture descriptions* and architecture description languages which are not discussed here in more detail. As the Smart Grid means increased use of ICT for enterprises, these foundations about architecture shall now be illustrated in the context of enterprises and Smart Grid models.

4.2.1 Enterprise Architecture in the Context of Smart Grids

The functions of enterprises in the Smart Grid context differ from country to country, mainly depending of the type of market and regulations. Examples of such roles, which are taken on by enterprises, are network operators, suppliers, traders, electricity generators or measurement service providers. Due to new roles and functionalities introduced with Smart Grids, enterprises are required to exchange various information with others and to adapt their business processes as well as their supporting information technology. A particular future challenge for distribution system operators for example is represented by the integration and management of their field/operational technologies. Such integration efforts are required and will become more common in context of the increasing amount of distributed generation. As already mentioned, new functionality can also mean new opportunities for the business, i.e. in terms of products or services. This provides a reason for enterprises to take more consideration of the efficient operation of their business- and IT in order to realize new opportunities quickly.

Enterprise architecture deals with the system "enterprise" and so with typical stakeholders and concerns within an enterprise and its environment. There are various definitions of the term "enterprise architecture" available, but there is no generally accepted definition. The understanding of "enterprise architecture" can vary from modeling the enterprise (in terms of architecture description), over implementing architectures within enterprises to a professional discipline or method to manage the enterprise architecture (architecting and managing the architecture). Regardless of the understanding, the subjects of "enterprise architecture" often regard artifacts from the business strategy, organizational issues, business and IT integration as well as software and technical infrastructure [1].

According to the definition given in [8], the term enterprise architecture is understood in the following as defined up to this point with the enterprise representing the system-of-interest. In the context of enterprises, which are socio-technical systems, it is especially important to align the information technology (IT) used (e.g. data entities or information system services) with the respective functions of the business (e.g. business goals, business processes) and organization in order to efficiently operate the business. This alignment process is a focus of enterprise architecture, which holistically considers the system "enterprise". Without a defined/documented enterprise architecture, it will be difficult to consider and meet the stakeholders' concerns and requirements. Viewpoints for enterprise architecture shall allow to express such concerns.

4.3 Viewpoints for Enterprise Architecture

There are several frameworks for enterprise architecture available, which provide different sets of viewpoints. In the late 80s, Zachman [15] provided first foundations for stakeholders and architecture viewpoints with his "framework for information systems architecture", which is often referred to in the context of enterprise architecture. He addressed the six basic concerns (what, how, where, who, when, why) for different stakeholders (strategists, executive leaders, architects, engineers, technicians, workers) and arranged them in a matrix where each cell is addressed by one or more viewpoints (which roughly conforms to the definition of an architecture framework). However, Zachman, providing a fixed set of viewpoints which are assumed to be complete, did not provide a process how to apply the framework. Basically, its application can lead much documentation and leaves it to the user to specify the model kinds and notations to use.

The Open Group Architecture Framework (TOGAF) [13] is a wide-spread and mature enterprise architecture framework which is elaborated by several large industry players as members of The Open Group. TOGAF provides three high-level architecture viewpoints for enterprise architecture, which are called "business architecture", "information systems architecture" and "technology architecture". The "information systems architecture" viewpoint is further subdivided into "application architecture" and "data architecture". According to [13] the following information artifacts and stakeholders are addressed by these viewpoints:

Business Architecture Business strategy, governance, organization, and key business processes information, as well as the interaction between these concepts are part of the business architecture. This viewpoint addresses the concerns of users, planners, and business management.

Data Architecture The structure of an organization's logical and physical data assets and data management resources. It addresses the concerns of database designers, database administrators, and system engineers.

Application Architecture A description of the major logical grouping of capabilities that manage the data objects necessary to process the data and support the business. Here, the concerns of system and software engineers are addressed.

Technology Architecture The logical software and hardware capabilities that are required to support deployment of business, data, and application services. This includes IT infrastructure, middleware, networks,communications, processing, and standards. Acquirers, operators, administrators, and managers are relevant stakeholders for this viewpoint.

TOGAF tries to incorporate the ISO 42010 as far as possible, but the word "architecture" here suggests, that there are four (independent) systems (business, data, application and technology) that have an architecture. In terms of ISO 42010, enterprise architecture is considered as a holistic conception, which can be observed from multiple viewpoints. These "architectures" are here understood as viewpoints, which frame respective concerns regarding business, applications, data, and technology, and so show decompositions of the enterprise.

Within each of these generic viewpoints, TOGAF identifies several exemplary viewpoints which can serve as a starting point to address particular concerns. These viewpoints are divided into three types: Catalogs, Matrices and Diagrams. Catalogs provide lists of information regarding architecture building blocks, matrices are to display relationships between them and diagrams are richer, graphical representations of these information and thus being more suited for stakeholder communication. More detailed "business architecture" viewpoints are for instance "Driver/-Goal/Objective Catalog" or "Actor/Role Matrix", more detailed "data architecture" viewpoints are "Data Entity/Data Component Catalog" or "Data Entity/Business Function Matrix". The TOGAF specification still recommends to take the stakeholder's concerns into account in order to create the architecture description. This also means, that not all of the proposed viewpoints may always be applicable and new ones may have to be developed to cover the concerns.

Not particularly enterprise-specific but Smart-Grid-specific architecture viewpoints are defined within the work of the EU Mandate M/490 CEN/CENELEC/ETSI Working Group "Reference Architecture" [3]. There, the so-called Smart Grid Architecture Model (SGAM) framework is defined, which allows a cross-domain localization of systems. This is a very important aspect in the context of Smart Grids, as it allows to identify interfaces between participating Smart Grid stakeholders. Thus it can enable interoperability between them which is a key to efficiently realize such a complex system. It consists of several layers representing a business viewpoint, a function viewpoint, an information viewpoint, a communication viewpoint and a component viewpoint. Each layer defines a matrix that allows the identification of a Smart Grid domain (one of Generation, Transmission, Distribution, Distributed Energy Resources (DER) or Customer Premise) and the identification of information management zones (one of Market, Enterprise, Operation, Station, Field or Process). The SGAM is described in more detail in the context of requirements engineering in Chapter 2.

The articulation of an enterprise architecture is the basis for its planning, management and evolution. To develop an enterprise architecture description—for instance for planning, management, or just documentation purposes — a well-structured method considering stakeholders, their needs and the organization of created artifacts is required. TOGAF provides these structures and a method for the development, implementation and management of enterprise architectures, independent of any particular business domain. The basics of this approach will be outlined in the following to provide foundations for the enterprise architecture management in the energy sector.

4.4 An Approach for Enterprise Architecture Development and Its Management

The definition of an architecture basically requires to break down a system into its parts, and then to proceed in the same way with its parts until a sufficient granularity for description depending on the objectives of the architecture is reached. This assumes hierarchically structured systems, which are composed of interrelated subsystems, i.e. "nearly decomposable systems" according to [11]. The decomposition of a particular problem area helps to make the complexity of large systems more manageable. Having only to consider parts of a system reduces the complexity of each part and in principle requires only to consider defined relationships. Particular parts of related functionality (e.g. in the form of building blocks) interacting with each other, realize the overall system. In addition, these building blocks provide a basis for work and resource planning, e.g., for scheduling and timing of work tasks, cost analysis or risk management.

One single enterprise already involves various aspects to be taken into account in enterprise architecture. Regarding the integration and information exchange across multiple enterprises with heterogeneous systems, the development can even become more complex. In order to align the changing business and IT-environments and their large numbers of systems, a continuous enterprise architecture management practice has to be established. Since Smart Grid concepts and unbundling provisions in the energy sector mean change to the business of several enterprises and require more information exchange, an established approach, like provided by TOGAF [13], seems reasonable. TOGAF provides methods and tools to develop, implement, use and maintain an enterprise architecture and also includes best practices in the form of a content framework as well as guidelines and techniques.

4.4.1 A Method for Enterprise Architecture Development

The central method of TOGAF is called Architecture Development Method (ADM). It provides a proven and repeatable process for developing architectures. The ADM defines ten phases which can be executed in different iterative cycles, continuously defining and realizing the architecture to a certain extent. Figure 4.4 depicts the phases of the ADM, which are adapted and described in brief in the following text.

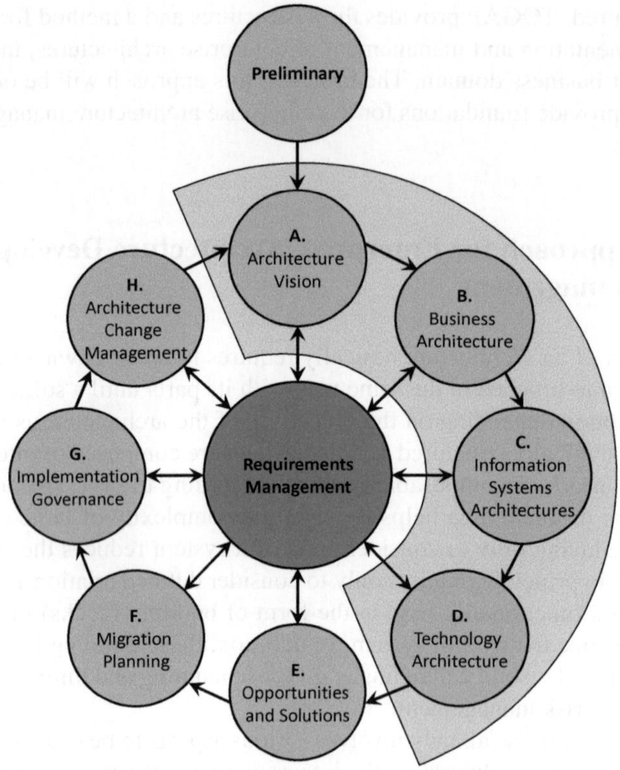

Fig. 4.4 The TOGAF Architecture Development Method (ADM) [13]

As architecture development is a quite generic task, the *Preliminary* phase is about the preparation of the architecture development within the enterprise, e.g., the definition of an enterprise-specific framework or setting up principles for development. The subsequent *Architecture Vision* is the first phase of an architecture development cycle. It is about the definition of the envisioned architecture scope, analysis of stakeholders and the definition of an initial outline of the target architecture addressing the architecture viewpoints (business, information systems and technology) which are in the scope of the development cycle.

Following the *A. Architecture Vision*, more detailed business, information systems and technology views are developed in the phases *B. Business Architecture*, *C. Information Systems Architecture* and *D. Technology Architecture* respectively. Again, the term architecture can here be understood ambiguously: on the one hand, it can be understood as a viewpoint framing concerns related to business, information systems and technology, which are viewpoints on the system "enterprise". On the other hand, it can refer to business, information systems or technology as systems which have an architecture. They identify elements (in a particular domain), their relations and principles and puts them into relation. When referring to one of these "architectures" in the following text, the architecture viewpoint or the development of the respective view is meant.

Primarily, the ADM follows a business-driven approach (top-down), i.e. business requirements are the driver for ICT in enterprises. This means, that information systems and technologies for the Smart Grid are solely required to accomplish business goals. Thus, the phase *B. Business Architecture* deals with identification business concerns and the development of the business architecture view. This comprises for instance capturing business processes and associated requirements, which have finally to be realized by information technology (hardware and software, phases C./D.).

However, it is also appropriate to begin with the identification of technology requirements and the definition of the technology or information architecture view (bottom-up), which is especially useful in case of emerging and technology-driven areas like the Smart Grid. Here, new, innovative technologies would be deployed whose additional value can affect several business areas. The deployment of Smart Meters could for instance influence consumer behavior and so affect elements in the business architecture in the form of changed business models (e.g., nearly real-time tariffs). In practice, a mix of both approaches—top-down and bottom-up—will be reasonable to exploit innovative potentials in the long run, and to optimally support the business with IT. For an economical operation, costs and benefits will have to be balanced against each other, especially in the context of the envisaged period of use and the generally short innovation cycles in the technology sector vs. the rather long innovation cycles in the context of business models.

The phases *Information System Architectures* and *Technology Architecture*, following *Business Architecture* phase, address the development of the actual elements in the scope of the architecture view. This comprises for instance the identification and development of business data entities and information system services in the *Information System Architectures* phase, and technology components or platform services in the *Technology Architecture* phase.

All of the phases B.–D. generally capture/document the current architecture (baseline architecture) and define an envisioned architecture (target architecture). This information provides the basis for a gap analysis to derive particular actions for its realization (transition architecture).

Following phase D., the ADM phase *E. Opportunities and Solutions* deals with the initial implementation planning of the previously defined architecture. This comprises the review of objectives and artifacts developed so far, their consolidation, consideration gap analysis results and the definition of how to deliver the architecture. Transition architectures are defined here to incrementally develop the architecture in several stages, still maintaining normal business operation.

Phase *F. Migration Planning* considers the formulation and coordination of a series of transition architectures, providing an implementation- and migration-plan. This includes among others, the prioritization (in terms of business value) of work packages, projects and building blocks, the finalization of architecture definition documents and the final confirmation of actions from relevant stakeholders.

The subsequent phase *G. Implementation Governance* addresses the governance of implementation projects, e.g. that the solutions meet the plan and architecture requirements. Within this phase, also the initiation of activities which are required for the operation of the implementation takes place.

In the last phase *H. Architecture Change Management* of the architecture development cycle, procedures for monitoring and reacting on changes to the new architecture are set up. These procedures shall ensure, that the new baseline architecture fulfills the requirements and when a new iteration of the ADM is to be triggered.

Finally, the central *Requirements Management* phase is related to all other phases, as illustrated in Figure 4.4. It is concerned with managing architecture requirements and making them accessible within the phases of the ADM. Requirements identified in the *Business Architecture* phase can for example have effects on the applications. Requirements identified in later phases can have effects on work done in previous phases and so the requirements management phase is to allow the consideration of this information.

TOGAF represents a framework and needs to be tailored for application by a particular enterprise. Regarding the ADM, iterations may for instance be carried out in other sequences if this fits better to the organizations goals. TOGAF itself also suggests more specific iteration cycles, e.g., an iteration cycle between *Preliminary* and *A. Architecture Vision* to define the architecture context, an *Architecture Definition Iteration* between *B. Business Architecture* and *F. Migration Planning* (including sub iterations for *C. Information System Architectures* and *E. Opportunities & Solutions* and *F. Migration Planning*), or an *Architecture Governance Iteration* between *G. Implementation Governance* and *H. Architecture Change Management*. Some of these cycles may for instance be executed once, others more often, which mainly depends on the scope and objectives of the development effort.

Additionally, the focus on baseline or target architectures can differ per iteration. Generally, the identification of baseline architectures will for instance be done in early iterations of the architecture definition iteration. In the context of Smart Grids, capturing and considering the current architecture is also an important step. Since it can be assumed that it will not be newly build from scratch, the use and integration of existing infrastructure is required and so its documentation is inevitable. Figure 4.5 shows the states between baseline and target architecture, outlining incremental architecture development. The target architecture represents an ideal to be reached. Oftentimes it is not possible to reach this envisioned state, e.g., resources required for realization are not available in terms of technology, restrictions in time, budget, or as changed requirements imply changes to the envisioned target. Transition

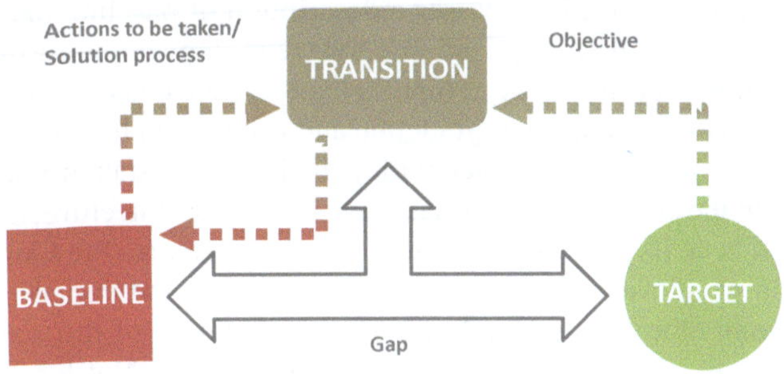

Fig. 4.5 Architecture development as a transition between baseline and target architecture

architectures are increments showing periods of transition and development for particular parts. They should be realized in specific projects and provide the basis for planning. They are defined in the last phases of an architecture definition iteration.

After the brief overview on the ADM, the phases most relevant for the architecture definition and requirements management shall be described in more detail.

4.4.2 Preliminary Phase

The preparations made within this phase are valid throughout the ADM-iteration. This includes to define why the development takes place, which stakeholders are involved, which scope of the enterprise architecture is considered and where and how the development process is conducted. As already mentioned, this phase can also be executed in parallel with or after phase *A. Architecture Vision* in terms of the *Architecture Content Iteration.*

Goals of this phase are to tailor the development process and to define in detail which methods are used to define the architecture. This can for instance mean to define the focus of the development effort within a cycle, as well as the individual activities to be done in particular phases. The last point also includes the definition of specific tasks, different work products, involved roles or individuals.

Another concern addressed in this phase is the definition of fundamental principles regarding the architecture (so called architecture principles). This can include the definition of enterprise principles or information technology principles. Additionally, tools to support the development process are identified, defined and introduced respectively.

4.4.3 Phase A: Architecture Vision

A first version of the envisioned architecture is developed in this phase. It determines the area and scope of the architecture development effort, defining the boundary of what is part of the target architecture and information which will not be considered.

The architecture vision encompasses a description of baseline and the target architectures for business, data, applications, and technology domains on a high-level. These architectures' outlines serve as input for subsequent phases which they are further developed. As it shall serve as guideline and foundation for the development, it should be widely-known and accepted within the identified stakeholder community.

In addition to the creation of an initial version of the architecture, this phase shall also outline the benefits for stakeholders. At first, this requires the identification and documentation of relevant stakeholders and their goals and concerns. This information is, among other things, the basis for the definition of the scope and focus of the architecture, which is also determined in this phase. Not least, this requires the identification and prioritization for further development of components of the baseline architecture, which can be used as input for the vision. This phase also targets to define essential business requirements which shall be addressed with the architecture development effort. In this context, existing business principles, goals and strategic drivers of the organization are to be identified. At the same time, this phase is also the beginning of a development cycle and therefore also includes motivational and organizational work. This comprises for instance the organization and definition of the development cycle within the boundaries defined in the *Preliminary* phase. Beyond that, planning of resources (time, finances, people), communication, risks, constraints, assumptions, and dependencies will be carried out in this phase.

4.4.4 Phase B: Business Architecture

Based on the *Architecture Vision*, phase B. elaborates the business architecture in more detail. The business architecture represents a viewpoint on the enterprise architecture. According to [13], it basically defines the business strategy, governance, organization and essential business processes. Business goals can be refined and decomposed to define business requirements and finally business services. These services inter-exchange data in the form of business objects, which are also identified in this phase. Additionally, business services can involve several roles which can also be identified. The business architecture provides input to the subsequent phases, defining its realization through IT.

However, defining the business architecture requires also to identify and examine appropriate architecture viewpoints, which include information relevant for the particular stakeholders. Furthermore, relevant tools and techniques supporting the development of respective views have to be selected. A gap analysis between baseline and target business architecture descriptions finally allows to derive further development actions.

4.4.5 Phase C: Information Systems Architectures

The *Information Systems Architectures* phase consists of two parts, considering different aspects of information systems: *Application Architecture* and *Data*

Architecture. Within this phase, baseline and target architectures for the data and application domain are being defined. Depending on the architecture project's objectives, the focus on only one of the domains, i.e. either data or application may be possible. Moreover, these two viewpoints and their respective views shall also illustrate relations to the business architecture in terms of a realization relationship. Precisely, this means, that information (represented by business objects) is expressed as data and business processes or their functionality respectively is realized by applications.

Furthermore, the order in which the relevant views are developed can vary and can be chosen depending on the context. There are data centered development approaches, starting with the data viewpoint, but also the functionality regarding the business processes to be realized can be used to argue for an application centered beginning.

On the one hand, the *Application Architecture* viewpoint identifies individual application systems including their relationships to the organization's core business processes (identified in the previous phase) they support. Thus, the viewpoint provides blueprints for these systems' development. The focus is not on the design of these systems, but rather on the identification of different system kinds and the requirements they address. It moreover depicts the interactions and relations to core business processes. Applications are understood as logical groups of functionality which process data objects of the data architecture and support business functions. The description of applications takes place on a logical level, i.e. technology-neutral. Identified applications are relatively persistent over time while technology changes more often. Typical applications in the energy sector are, e.g., Supervisory Control and Data Acquisition (SCADA) systems, although the description of their functionality is quite abstract. Further applications may for instance deal with billing, master data management or planning.

On the other hand, the *Data Architecture* includes the structure of logical and physical data required to exchange information and data management resources. It is concerned with the identification of high-level, enterprise-wide data, i.e. information relevant for business processes, but not with the design or development of data management systems like databases.

This is especially relevant for the understanding of data management, the migration of data, their maintenance and also their quality management. Possible considerations regarding data management can for instance begin with the identification of components relevant for creation, saving or use of data.

In the context of Smart Grids, the information collected in this phase is for instance required in order to exchange customer data, metering data or billing data across several involved Smart Grid actors. Within the energy sector, the *Common Information Model (CIM)*, defined in the IEC standard 61968/61970, is widely used. As a well-elaborated approach it is strongly recommended to incorporate this model in respect of strategic orientation.

4.4.6 Phase D: Technology Architecture

According to the previous phases, a baseline and target architectures are defined based on the identified business goals. The technology architecture is to define the physical aspects of the realization of the information systems architectures and hence the business architecture.

Considering a "complete" realization of the Smart Grid, it has to be determined to which extent the technology architecture shall be developed. Depending on the enterprise's market role, operational technology located in the field, like digital meters, home gateways or substation automation technology, can for instance be considered as "in scope" or be explicitly excluded. However, where the boundary of an architecture is drawn of course depends on the scope of the specific enterprise architecture effort. Through the increasing use of modern ICT, the consideration of operational technologies may yield synergy effects.

Classically, the technology architecture viewpoint describes logical hardware and software capabilities required to provide business, data and application services in the context of the enterprise architecture. As far as these components are involved in the delivery of these functions, they are part of the technology architecture. Again, depending on the scope, only abstractions of these technologies may be sufficient and, for instance, result in the identification of needed standards.

Elements of the technology architecture are, among others, IT-infrastructure, middleware, networks, and communication standards. In the context of standards, it is recommended to incorporate the standards framework given in IEC TR 62357. Further, aspects of legislation and regulation are to be taken into account, as they may prescribe specific technologies. By its character, the technology architecture provides links to implementation and also migration. These tasks is dealt with in the subsequent phases, which are not in the further scope of this chapter.

4.4.7 Requirements Management Phase

Another important task is requirements engineering, which is key to all phases in the TOGAF ADM. Especially in the context of Smart Grids, requirements are not clear or fix, but rather highly dynamic. Thus, requirements and changes to the requirements occur and must be tracked and their impacts be analyzed in all phases. That means, that all requirements must be captured centrally in terms of a defined requirements management and made available to other phases. Requirements are only collected but not prioritized, which is part of the respective phases. The process or documentation for the requirements management phase is not prescribed by TOGAF. For details on requirements engineering approaches in context of Smart Grids, therefore please refer to Chapter 2. Also use cases have proven useful in regarding the elicitation and documentation of requirements. A recommended, energy sector specific method is provided by the IEC specification IEC/PAS 62559 [6]. A use case development and management methodology based on the IEC/PAS 62559 is provided in Chapter 3.

4.5 Conclusion and Outlook

The term *architecture* is often used with different meanings and purposes in mind. Architectures represent abstractions of systems and generally encompass elements and relationships of a system. Each system has an architecture, which is not always documented or explicitly visible. Architecture descriptions try to express systems's architectures from several viewpoints, depending on the architecture stakeholders and their concerns.

A structured development effort can help to address these concerns in the complexity arising with system such as the Smart Grid. One important part of "architecture development" is the definition of architecture descriptions, which is the basis for communication about systems' elements, their planning, implementation or evolution. These descriptions can for instance capture the system's structure or its behavior. Architecture descriptions result in models of the architecture defined with a specific purpose reducing the complexity of the architecture as a whole. These models again can be used define systems or structures thereof in a model-based way. Moreover, architecture models covering different states (e.g., target and baseline) enable analyses and migration planning.

The definition of architectures, i.e. architecture descriptions, is a non-trivial task and is usually carried out in a well-structured process like the TOGAF ADM in the context of an enterprise as the system-of-interest. As architectures are often developed as an envisioned state, respective implementation and governance is required.

In the context of the Smart Grid, there exist various sub-systems having an architecture, ranging from software systems, to hardware, or socio-economic systems like enterprises. However, in order to align these systems in order to inter-operate, well-developed systems are a desirable goal. Architectures can provide helpful abstractions here to define the scope of systems, specifying their elements and relationships.

While foundations regarding architecture development and management are, among others, provided by ISO/IEC/IEEE 42010 [8] or TOGAF [13], their application and tailoring for actors within the power system domain is subject to ongoing research (for more details, see [14]). Figure 4.6 outlines the authors' approach of classification and integration of methodologies and tools for Smart Grid-oriented enterprise architecture development. It shall be briefly described in the following as it may also be used to identify topics of interest related to architecture development within other chapters of this book. In the figure, the need for meaningful orchestration of systems owned by different actors is expressed by the *External viewpoint*. *Artifacts* like reference architectures (e.g., the SGAM mentioned earlier in this chapter), roadmaps concerned with standardization (e.g., [4]) or technology (e.g,. [2], discussed in context of Requirements Engineering in Chapter 2), regulatory provisions and shared use cases (see Chapter 3) may among others be of interest as sources of requirements from the external viewpoint. Existing methodologies (e.g., SGAM or IEC/PAS 62559) regarding these artifacts may be used to gain access to this information. Regarding the application of shared use cases there additionally exists tool support in form of an Use Case Management Repository (UCMR).

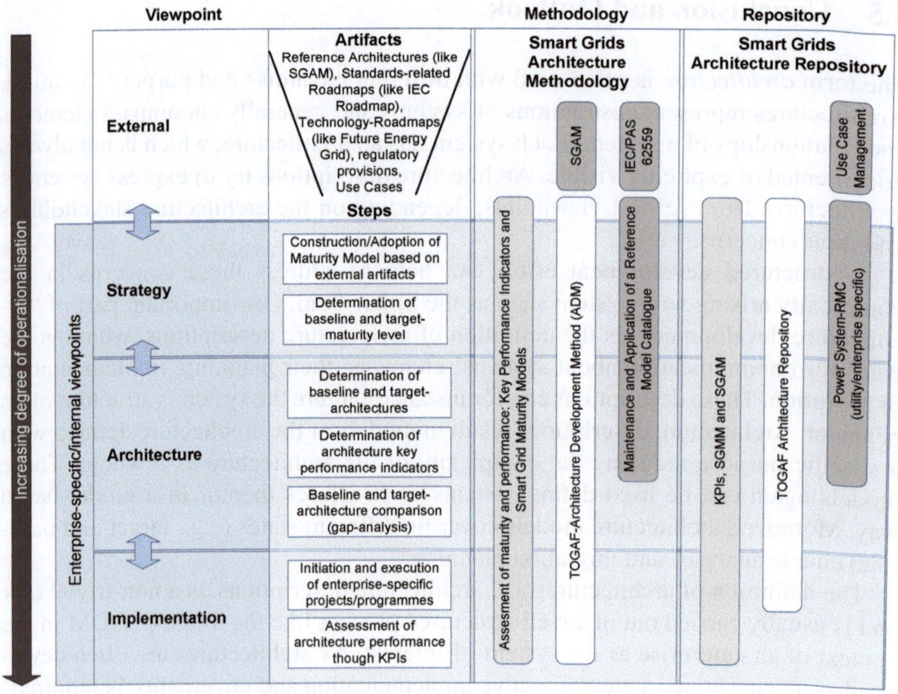

Fig. 4.6 Proposed integration of methodologies and tools regarding Smart Grid-oriented enterprise architecture development

Using a top-down approach, the enterprise's *Strategy* regarding Smart Grid programs represents the first *Enterprise-specific viewpoint*. We propose the construction of a maturity model in order to identify and assess maturity levels regarding technologies, products or business processes relevant to the enterprise in context of Smart Grid implementation. By determining the baseline and target maturity levels the enterprise is enabled to establish a strategic assessment tool. TOGAF as discussed in this chapter may for example integrate the maturity model in context of the architecture vision step of the ADM and use it in context of the migration planning phase to identify suitable transition architectures. The TOGAF Architecture Repository (see [13]) and the application of a Power-System-specific Reference Model Catalogue (as discussed in Chapter 5) provide the means to structure and preserve the information needed in this process.

Based on knowledge gained from the Strategy view, baseline and target architectures can be derived in the context of the *Architecture* viewpoint. The levels of maturity identified in context of the Strategy viewpoint are refined into architecture key performance indicators (KPI). These provide impartial criteria to analyze the gaps between baseline and target architecture and shall be used in context of the *Implementation* viewpoint to assess the current architectures performance.

Like in the ADM, this holistic perspective on enterprise architecture development should be elaborated iteratively with respect to the interdependencies between the viewpoints and their views' elements. While architecture descriptions represent an abstraction from both, business and implementation needs, valuable requirements originate from these perspectives.

References

1. Aier, S., Riege, C., Winter, R.: Unternehmensarchitektur Literaturüberblick und Stand der Praxis. Wirtschaftsinformatik 50(4), 292–304 (2008)
2. Appelrath, H.J., Kagermann, H., Mayer, C. (eds.): Future Energy Grid - Migrationspfade ins Internet der Energie. Springer, Heidelberg (2012)
3. Bruinenberg, J., Colton, L., Darmois, E., Dorn, J., Doyle, J., Elloumi, O., Englert, H., Forbes, R., Heiles, J., Hermans, P., Kuhnert, J., Rumph, F.J., Uslar, M., Wetterwald, P.: Smart Grid Coordination Group Technical Report Reference Architecture for the Smart Grid Version 1.0 (DRAFT) 2012-03-02. Tech. rep., CEN, CENELEC, ETSI (2012)
4. DKE: The German Standardization Roadmap E-Energy/Smart Grid. VDE (2010)
5. IEC: 62357 Second Edition: TC 57 Architecture - Part 1: Reference Architecture for TC 57 - Draft (2009)
6. International Electrotechnical Commission (IEC): Publicly Available Specification (PAS) 62559 IntelliGrid Methodology for Developing Requirements for Energy Systems. Tech. Rep. 1.0 (2008)
7. ISO/IEC: ISO/IEC 10746-1:1998 Information technology – Open distributed processing – Reference model: Overview. Tech. rep. (1998)
8. ISO/IEC/IEEE: ISO/IEC/IEEE 42010 ed1.0: Systems and software engineering - Architecture description (2011)
9. Kruchten, P.: Architectural Blueprints — The 4+1 View Model of Software Architecture. IEEE Software 12(6), 42–50 (1995)
10. NIST: NIST Framework and Roadmap for Smart Grid Interoperability Standards (2010)
11. Simon, H.A.: The architecture of complexity. Proceedings of the Americal Philosophical Society 106(6), 467–482 (1962)
12. Smolander, K.: Four Metaphors of Architecture in Software Organizations: Finding out The Meaning of Architecture in Practice. In: Proceedings International Symposium on Empirical Software Engineering, pp. 211–221 (2002)
13. The Open Group: TOGAF Version 9 - The Open Group Architecture Framework (TOGAF), 9 edn. (2009)
14. Trefke, J., Dänekas, C., Rohjans, S., González, J.M.: Adaptive Architecture Development for Smart Grids Based on Integrated Building Blocks. In: 3rd IEEE PES Innovative Smart Grid Technologies (ISGT) Europe Conference (2012)
15. Zachman, J.: A framework for information systems architecture. IBM Systems Journal 26(3), 276–292 (1987)

Chapter 5
Management of Information Models in the Energy Sector

José M. González and Jörn Trefke

Abstract. This chapter motivates the use of information models to support the functional development of IT-applications for utilities. The complexity of the relationship between technology, regulation, business models and existing infrastructure in the domain of Smart Grids can be seen as very high. Other domains have adopted the concept of reference models to manage this complexity. This chapter provides an introduction to information models and presents a reference model catalog for the energy sector developed at OFFIS. It was developed with a focus on multi-utility enterprises in Germany but could also be extended to other domains and geographical regions.

5.1 Smart Grids and Challenges for Enterprises

The German energy industry is undergoing a process of structural changes due to changing regulations and technical advancements, see e.g. [4], [17] or [14]. On the one hand laws have been approved to encourage competition in the German energy sector like the legal unbundling as described in the German energy industry act (Energiewirtschaftsgesetz (EnWG) [7]). On the other hand technical advancements lead to new products and services like Demand Side Management (DSM) and Automated Meter Reading (AMR). With the upcoming distributed generation, the legal requirements imposed by federal regulation and the resulting unbundling, the situation has changed to a large extent. Due to new generation facilities, like wind power plants or fuel cells, energy is fed into the grid at different voltage levels and by different producers—former customers having their own generation can now both act as consumers and producers (also referred to as prosumer) which feed into the utilities' grid. Therefore, the communication infrastructure has to change.

José M. González · Jörn Trefke
OFFIS – Institute for Information Technology, Escherweg 2, 26121 Oldenburg, Germany
e-mail: {jose.gonzalez, trefke}@offis.de

M. Uslar et al.: *Standardization in Smart Grids*, POWSYS, pp. 79–96.
DOI: 10.1007/978-3-642-34916-4_5

The energy sector comprises several activities like generation of electricity, gas, fuel or district heating. To reduce the complexity within this chapter, we only address electricity and gas when referring to the *energy sector*, as a major part of the German energy sector (45 % of the energy consumption) [?]. In addition, electricity and gas have (with regard to business transactions) several processes in common despite of their physical differences.

Current application landscapes for utility companies were built to address requirements of the past de facto monopoly environment. Today, companies in the energy industry face more competition and have to provide new products and services at lower costs. This requires current application landscapes to become more flexible and to be able to adapt faster to the evolving requirements resulting in structural business changes. Therefore, adequate IT-infrastructures supported by appropriate architectures, like service-oriented architectures, are needed [22]. Both utility companies as well as software manufactures have to deal with these changes and need to adapt their application landscapes or software products. In this context, requirements analysis plays an important part.

Current national and international initiatives (like E-Energy[1] and the European Technology Platform on Smart Grids[2] respectively) and discussions in the energy sector reveal that the power system is developing towards a so called "Smart Grid". The Smart Grid vision encompasses the integration of multiple devices and actors continuously exchanging data to provide user-oriented flexible services and products while operating a self-healing, economic, ecologically friendly and secure network.

All these changes lead to new requirements regarding business information systems which support core tasks and processes of enterprises in the energy sector, e.g. additional functionalities or new IT security requirements have to be provided/-considered. Software product managers in energy sector and software developing companies who are in charge of driving the functional development of information systems have to deal with those challenges and need to develop new information systems or enhance existing ones.

Due to current technical and organizational changes an increasing number of heterogeneous information sources need to be taken into account within the requirements analysis. Corresponding information sources grouped by general, regulatory and technological influencing factors are outlined in Figure 5.1. In addition, Figure 5.1 presents traditional activities of enterprises in the energy sector, which are based on two interacting supply chains. These supply chains focus on the one hand on the cash flow (Business viewpoint), and on the other hand on the power or gas flow (Engineering viewpoint). Based on this, typical market roles of enterprises in the energy sector are listed. As the transport (transmission and distribution) of power and gas is subject of regulation (natural monopoly) the supply chain element is highlighted. Finally requirements of enterprises and their information technology are outlined.

[1] `http://www.e-energy.de`

[2] `http://www.smartgrids.eu`

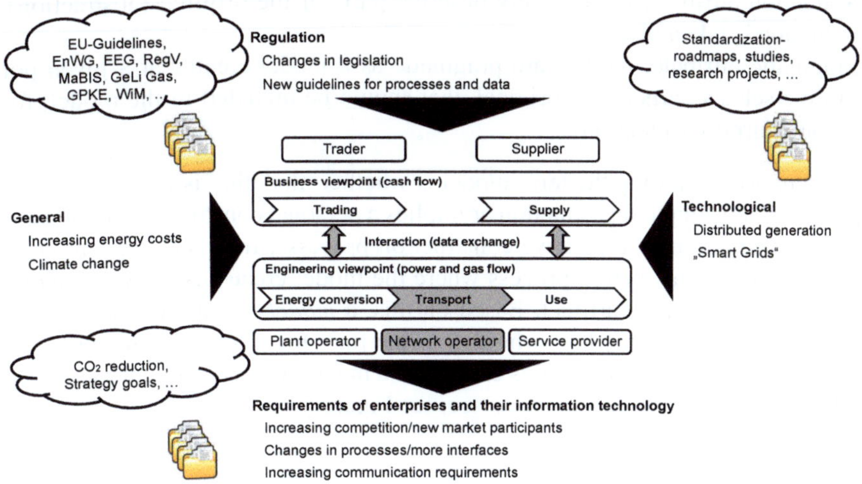

Fig. 5.1 Influencing factors for structural changes in the energy sector [12]

The Smart Grid requires the application of standards for being able to cope with heterogeneity and enable interoperability in an economic and technically feasible way. In addition, existing knowledge, often described in (functional) reference models or standards, should be used to design efficient processes and identify required functionality. Therefore the remainder of this chapter is organized as follows. After outlining the multitude of challenges the energy sector and its companies have to deal with, an introduction to information models (Section 5.2) presenting several types of information models, as well as an approach for their classification is provided. Next relevant information sources for requirements analysis (Section 5.3) and information systems (Section 5.4) are outlined. In addition, Section 5.5 introduces the energy reference model catalog as research approach to structure the multitude of information sources. Finally Section 5.6 and provides a summary and an outlook on future work.

5.2 Introduction to Information Models

In the field of information systems development, conceptual (information) modeling has been known for many years. It is often applied to analyze, design, and implement information systems. Even though modeling is done for years the definition of the term model is still subject of discussions, see e.g. [28], [30] and [12].

According to Stachowiak [26] a model is characterized by the following three attributes:

1. Mapping attribute: a model always represents an original to which it defines a mapping relation (reference).

2. Reduction attribute: models only describe parts of the original (abstraction), a reduction is taken place.
3. Pragmatic attribute: models are pragmatic as a model creator at a given point in time selects parts of the original, that should be included in the model, for a certain purpose (intention).

One discussion regarding the term model is whether modeling is only the result of observation and simple reproduction or implies a personal contribution of the model creator and therefore is a creative construction process[3]. In this chapter modeling is regarded as a construction process where the model creator is actively involved and develops a valuable artifact. Based on this, a *model* is regarded according to Steinmueller [27] as a "model – whereof – what for – for whom". Following [27], a model should be defined within this chapter as a mapping of an original by a subject (model creator) to influence an addressee, see Figure 5.2.

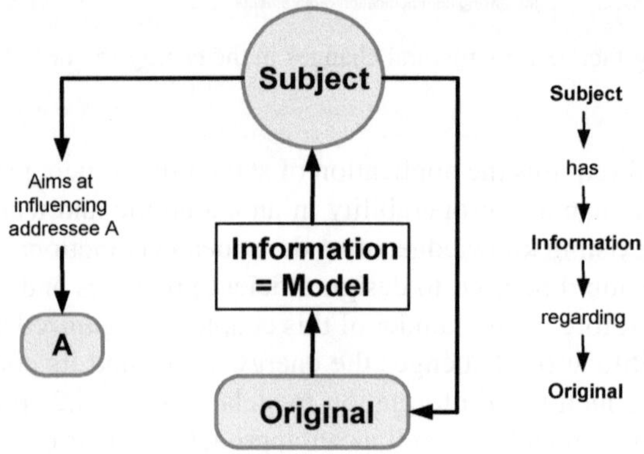

Fig. 5.2 The term model according to Steinmueller [27]

Regarding the purpose of modeling several goals exist. In literature, e.g., the representation of a relevant part of reality or an original, providing a common basis for communication (for decision makers) or to support the development of software or organizations and processes are identified. Due to the different purposes various model types like process or data models are available. This chapter deals with the support of business information systems in the energy sector and especially focuses on information models and excludes the interrelated information systems and organization design. An information model is defined as a "specific model which provides valuable information for information systems and organizational development" [25].

[3] See, e.g., [28], [30] and [12] for further definitions and discussions regarding the term model.

5.2.1 *Types of Information Models*

In Table 5.1 a morphological box of information model types is provided to illustrate the several types of information models that are available. Table 5.1 outlines several attributes and corresponding values grouped by the following categories: *model aspects*, *modeling language*, *support for reuse*, *technology* and *access*. *Model aspects* describe the purpose and coverage of the information model, *modeling language* defines the type of modeling language used, *support for reuse* species supported techniques to apply the information model), *technology* describes the kind of provision and *access* outlines possibilities of access.

Table 5.1 Morphological box regarding the classification of information models on [30], [25], [33], [10] and [1]

Attribute	Value			
Model aspects				
Intention [1]	As-is information model	Target information model		Ideal model
Content [30] [1]	Enterprise-specific information model		Reference information model	
Purpose perspective [30] [33]	Organizational model		Business information system model	
Layers [30], [33], [1],[25] [24]	Business concept	Data processing concept	Implementation	
Viewpoints [33] and [10]	Viewpoint-specific			Viewpoint overlapping
	Structure	Behavior Function	Extended model	
Views based on [1]	Data	Functions Information systems	Roles Products	Processes Maturity
Degree of aggregation [30]	Micro		Macro	
Abstraction layer [25]	Instantiation	Type	Meta	Meta-Meta
Type [25] (see also [15] and [21])	Object information model	Meta information-model	Meta^n information-model	
Task type [33]	Support		Core	Management
Industry sector [33] and [10]	Electricity	Gas	Trading	...
Enterprise function [10]	R & D Production Accounting	Sales Logistic Human Resources	Procurement Customer Care Facility management	Warehousing Finance Others
Modeling language				
Language [10]	EPC	OO (object-orientation) (UML)	Function tree	BPMN
Formality [33]	Formal	Semi-formal		Informal
Representation [33]	Graphical		Textual	
Extensibility [33]	Not extensible	Controlled extensible	Free extensible	
Support for reuse (especially for reference models)				
Construction techniques [5]	Configuration Analogy	Instantiation	Aggregation	Specialization
Amount of application models so far	None	One	> One	
Technology				
Representation [33]	Print		Electronic media	
Access				
Availability [33]	Not published		Partly published	Published
Provision	Navigable model		Editable model	Readable model
Legend	(shaded) Typical values of reference models according to [33]			

5.2.2 *Reference Information Models*

As the process of modeling is in general time-consuming and faulty, the concept of reference modeling was introduced. Here, reference modeling provides blueprints to improve and accelerate the modeling process. In addition, it aims at reducing modeling risks and costs to prevent failure of modeling projects. In this regard, model quality is considered as one major issue in reference model development. The term reference model is not clearly defined in literature, see e.g., analyses in [24], [33], [30], [29], [31] and [9]. Therefore several reference models exist for different stakeholders and purposes, like the Y-CIM [24] or the Zachman framework [36]. Within this contribution, a reference model is regarded as a blueprint that can be used to create a specific model in the context of information systems development or evolution. Therefore, a reference model is not regarded as an attribute of a model but as a relationship between two models [34].

Figure 5.3 illustrates the relationships between reference models and specific (or application) models based on a process oriented presentation. For the construction of reference and specific models the roles creator and user need to be distinguished. The creator is in charge of the development of the model according to the requirements of the user. Characteristic for the reference modeling is the consideration of different usage scenarios in advance ("Design for Reuse"). A reference model may serve as basis for several application models and hence offer efficiency benefits. In this case a new model does not need to be created from scratch. Instead, information already included in the reference model (and probably also already validated) can be used ("Design with Reuse") which might lead to cost and time reductions as well as increased quality. Through the use of reference models the development of high quality application models can be supported.

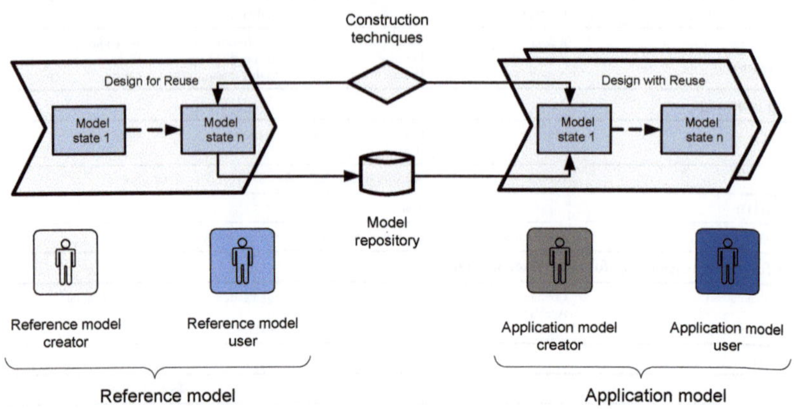

Fig. 5.3 Relationships of reference and application models based on [34]

5.2.3 *Reference Model Catalog Concept*

The increasing number of reference models leads to difficulties regarding the identification and selection of adequate and relevant models. Fettke and Loos developed the concept of a reference model catalog (RMC) aiming at providing a systematic and structured overview on reference models. In literature, the terms overview of reference models and reference model catalog are used synonymous. According to Fettke and Loos [10], a reference model catalog is typically structured as a table composed of three columns, see table 5.2. The first column provides a structure for classifying reference models (*Structure* part), the second contains the model names and authors (*Main* part) and the third one lists attributes, like modeling language, of the classified reference models (*Access* part). Several catalogs exist for different purposes and sectors, where some also include other sources than reference models, see [9]. This seems reasonable as the term reference model is not precisely defined and other information sources provide valuable information, too.

Table 5.2 Exemplarily table-based presentation of a reference model catalog based on [10]

Structure	**Main**		**Access**		
Industry sector	**No.**	**Author**	**Language**		
			EPC	ERM	Function tree
Manufacturing	1	Kurbel		X	
	2	Scheer	X	X	X
Retail	3	Becker, Schuette	X	X	X
...	...	...	X		X

5.3 Information Sources for Requirements Analysis within the German Energy Sector

The energy market offers a number of heterogeneous information sources which contain valuable domain knowledge regarding the development of information systems for the energy sector. Information sources (in the following referenced only as "sources") are all the documents which may or must be used in the development of energy sector-specific information systems, e.g., statutes, (reference) models, regulations, IT standards or ontologies. On the basis of desk research and expert interviews over 130 sources (like regulations, standards, and models) were identified in [12]. Here the identified sources are only exemplarily introduced, for a list of the identified sources and further details see [12]. These can or must (in case of regulatory demands) be used or considered in the requirements analysis for further development of application systems.

Figure 5.4 outlines an excerpt of the identified models (rectangle) and standards (rectangle with rounded corners) using the proven ARIS layers and views (see [24]) for classification. Hereby the penetration of standards regarding the ARIS-layers (rows, business to implementation with increasing level of IT topics) and -views (columns, organization to functions) is shown.

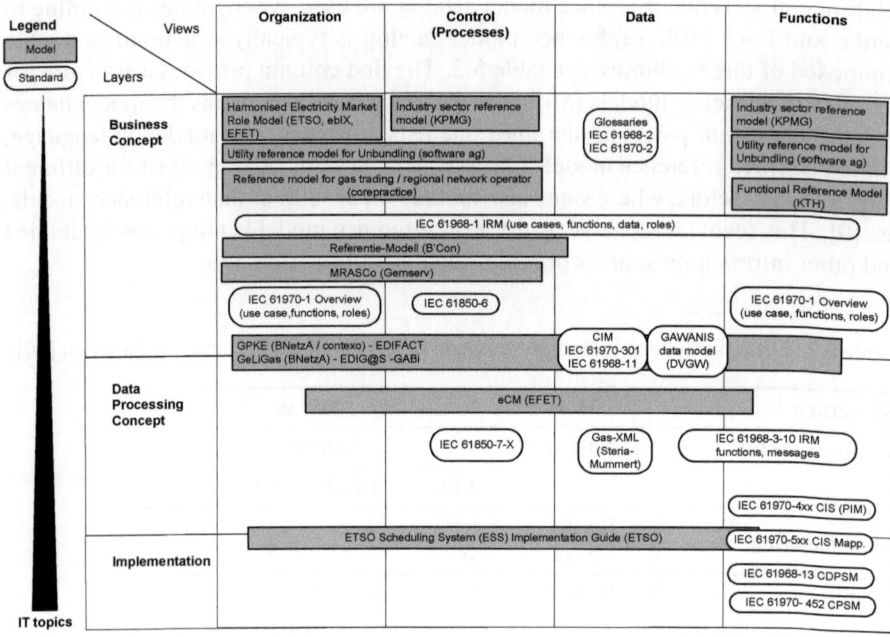

Fig. 5.4 Models and standards

The identified sources can be grouped into four categories: regulatory specifications, standards, (reference) models and ontologies. For the sake of clarity sources are only assigned to one category even though they may belong to more than one like for example the IEC CIM (see Chapter 6), which is a standard and a reference model. All models that were developed to be used as reference are considered here as reference models.

Legislators and regulation offices issue *laws*, *statutes*, and *regulations*. Examples hereof include the energy industry act (EnWG) as well as the rules for processes and data formats for data exchange between market participants (e.g., energy supplier and transmission system operator) for the supply of clients with electricity (GPKE) and gas (GeLi Gas).

Regarding Smart Grids and the required integration and control of market participants, applications and hardware, the international *standards* of the International Electrotechnical Commission (IEC) are relevant to develop interoperable solutions. Here, those of the Technical Committee 57 (TC 57) "Power Systems management and associated information exchange" are, according to a recommendation of the

"Deutsche Kommission Elektrotechnik Elektronik Informationstechnik" (DKE), of major importance. Especially relevant are the standard families IEC 61970 and 61968 which describe a data model for the integration of application systems of the energy industry. An overview on relevant Smart Grids standards is presented in Figure 5.5. The figure shows a list of exemplary standards which were identified and mapped to SGAM domains (see Chapter 2), the Energy RMC supply chain (see Section 5.5), and the TC 57 reference architecture (IEC TR 62357).

Reference models, often entitled as reference model or reference architecture by their creators, are being developed in collaboration with research institutes, associations, software producers, or consulting companies. Examples are the "functional reference model" of the KTH [20], the ENTSO-E "Harmonised Electricity Role Model" [8], the "Utilities Business Maps" [23] by SAP, or the "EVU-Referenzmodell" (Utility reference model) [18] of Software AG (formerly IDS Scheer).

In addition to that, several *ontologies* exist like the IT-security ontology of the Enertrust concept [2] and the E-Energy ontology of the German national electrotechnical standardization organization DKE.

In addition *glossaries* can be used for retrieving definitions and descriptions of core terms of the energy sector. Several glossaries are available online like the "IEC Electropedia"[4]) or the DKE E-Energy glossary[5].

To sum up, there are a multitude of heterogeneous sources like (reference) models, regulations, specifications, and further IT-technical or professional descriptions. These address various viewpoints (e.g., regarding data or functions) and target different development levels (e.g., requirements specification, design specification or implementation). They differ strongly in terminology, coverage and details regarding the various areas of value creation in the energy industry. This results from differing objectives, targeted stakeholders, and competencies of the sources' producers. In addition, sources undergo continuous changes and new ones are frequently added.

5.4 Information Systems in the Energy Sector

The complex tasks in the energy sector require an increasing use of information and communication technologies (ICT). ICT are gaining more and more importance in the energy sector as they allow for an efficient and economic operation of the different equipments and plants or even enable their operation. In particular, information systems—as a part of ICT—are used by utility enterprises throughout their whole supply chain supporting or even enabling their business processes.

Due to the varying use and enterprise-specific characteristics of information systems in the energy sector, a clear classification is not possible. However, with regard to activities of companies in the energy sector, technical (or engineering) and business activities are differentiated, see [16]. On this basis, information systems

[4] `http://www.electropedia.org`

[5] `https://teamwork.dke.de/specials/7/Wiki-Seiten/Homepage.aspx`

Standards		SGAM Domains					Criteria: Supply Chain								TC 57 Reference Architecture				
Standard	Description	Bulk Generation	Transmission	Distribution	DER	Customer Premise	Generation	Energy-Trading	Supply	Transmission	Storage	Distribution	Metering	Usage	Integration of business partners	Integration of applications	Integration of devices and plants	Security	Data management
AMI-SEC System Security Requirements	Advanced metering infrastructure (AMI) and SG end-to-end security			X									X			X	X	X	X
ANSI C12 Suite : (C12.1, C12-18, C12-19/MC1219, C12-20, C12-21/IEEE P1702/MC1221, C12.23, C12.24)	Revenue Meter Information Model			X		X							X	X					
BACnet ANSI ASH-RAE 135-2008/ISO 16484-5	Building automation					X								X			X		
Digital Meter/Homegateway	See EU Mandate M/441			X		X							X	X			X		
DNP3	Substation and feeder device automation			X								X					X		
EDIXML	Market Comunication with slow transition from EDIFACT to new CIM-bases technologies							X	X						X	X			
IEC 60870	Established comunication protocol		X	X						X		X					X		
IEC 60870-5	Telecontrol, EMS, DMS, DA, SA		X	X						X		X					X		
IEC 60870-6 / TASE.2	Inter-control center communications TASE.2 Inter Control Center Communication EMS, DMS		X	X						X		X					X		
IEC 61334	DLMS			X									X				X		
IEC 61400-25	Wind Power Communication EMS, DMS, DER	X			X		X										X		
IEC 61499	PLC and automation, profile for IEC 61850																		
IEC 61850 Suite	Substation automation and protection, DER, windfarms, hydro power plants, e-mobility	X	X	X	X	X	X	X	X	X	X	X	X	X			X		
IEC 61850-7-410	Hydro Energy Communication EMS, DMS, DA, SA, DER	X			X		X										X		
IEC 61850-7-420	Distributed Energy Communication DMS, DA, SA, DER, EMS	X			X		X										X		
IEC 61851	EV-Communication Smart Home, e-Mobility					X								X			X		
IEC 61968	Distribution Management, System Interfaces for Distribution Management Systems, DCIM (CIM for Distribution)			X								X				X			
IEC 61968/61970	Application level energy management system interfaces, CIM (Common Information Model) , Domain Ontology, Interfaces, Data exchange formats, Profiles, Process blueprints, CIM (Common Information Model) EMS, DMS, DA, SA, DER, AMI, DR, E-Storage	X	X	X	X	X	X	X	X	X		X	X	X	X	X			
IEC 61970	Energy Management, Application level energy management system interfaces, Core CIM	X	X				X			X						X			
IEC 62051-54/58-59	Metering Standards DMS, DER, AMI, DR, Smart Home, E-Storage, E-Mobility				X														
IEC 62056	COSEM DMS, DER, AMI, DR, Smart Home, E-Storage, E-Mobility			X	X	X						X	X	X			X		
IEC 62325	Market comunication using CIM							X	X						X	X			
IEC 62351	Security, Information security for power system control operations, security profiles	X	X	X	X	X	X	X	X	X	X	X	X	X				X	
IEC 62357	IEC 62357 Reference Architecture – Service-oriented Architecture, EMS, DMS, Metering, Security, Energy Management Systems, Distribution management Systems	X	X	X	X	X	X	X	X	X		X	X	X	X	X	X	X	X
IEC 62443 (ISA 99)	Method to achieve IT-Security regarding industry automation and control systems	X	X	X	X	X	X	X	X	X	X	X	X	X				X	
IEC 62541	OPC UA (Automation Architecture)					X								X		X	X	X	X
IEC PAS 62559	Requirements development method covers all applications.																		
IEEE 1547	Physical and electrical inter-connections between utility and distributed gen-eration (DG)	X					X										X		
IEEE 1686-2007	Security for intelligent electronic devices (IEDs)																	X	
IEEE C37.118-2005	This standard defines phasor measurement unit (PMU) performance specifications and communications for synchrophasor data.																		
ISO / IEC 14543	KNX, BUS					X								X			X		
MultiSpeak	A specification for application software integration within the utility operations domain; a candidate for use in an Enterprise Service Bus.																		
NERC CIP 002-009	Cyber security standards for the bulk power system	X	X	X	X	X	X	X	X	X	X	X	X	X				X	
NIST Special Publication (SP) 800-53, NIST SP 800-82	Cyber security standards and guidelines for federal information systems, including those for the bulk power system																	X	
Open Automated Demand Response (Open	Price responsive and direct load control			X		X							X	X			X		
OpenHAN	Home Area Network device communication, measurement, and control			X		X							X	X			X		
The Open Group Architecture Framework (TOGAF)	TOGAF is a framework, containing a detailed method and a set of supporting tools for developing an enterprise architecture.																		
ZigBee/HomePlug Smart Energy Profile	Home Area Network (HAN) Device Communications and Information Model					X								X			X		
Z-wave	A wireless mesh networking protocol for home area networks.					X								X					

Fig. 5.5 Relevant standards for Smart Grids

can be classified according the supported activities within the supply chains for electricity and gas: business, coordination or engineering (see Figure 5.6).

In the following, the individual information system categories are exemplarily described—a more detailed description can be found in [12]:

- *Business information systems* primarily aim at supporting functions related to the procurement and sales of energy (like electricity and gas) as well as related products and services. This comprises for example customer relationship management, billing or trading systems.

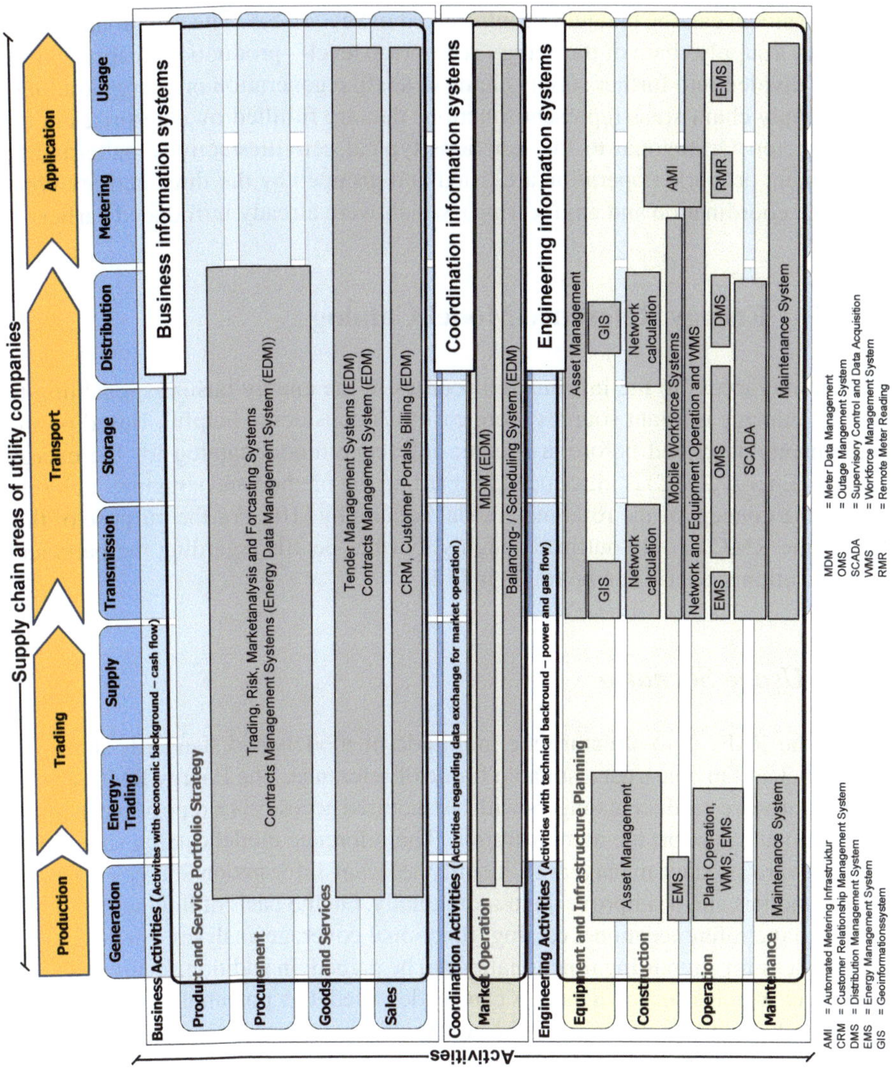

Fig. 5.6 Information systems in the energy sector [12]

- *Coordination information systems* focus on the interaction of the business and the engineering part and hence related information exchange between market participants. In this context meter data management systems, energy scheduling systems or balance management systems might be used.
- *Engineering information systems* deal with planning, construction, operating or maintaining of equipments like power plants, network and metering infrastructure, as well as distributed energy resources. Here for instance energy management, maintenance or outage management systems are used.

Figure 5.6 presents a functional matrix containing supply chain and activity areas containing exemplarily information systems. This matrix is the core element of the reference model catalog that will be introduced in Section 5.5. The functional matrix comprises a supply chain of the energy sector (top level—production to application) which is divided into further supply chain areas (like generation or energy-trading). These supply chain areas represent functions that are fulfilled by enterprises in the energy sector. Orthogonal to the functions typical activities/activity patterns like procurement or market operation are listed and grouped by the three activity types (business, coordination and engineering), which were already introduced earlier.

5.5 The Energy Reference Model Catalog

A structured access to the information sources in the energy business, enabling to find and manage relevant sources more easily, is considered helpful. Based on the requirements described before, a specific reference model catalog for the energy industry (Energy RMC) is introduced. Regarding the method and structure it closely follows the concept of the reference model catalog by [10]. For the purpose of this chapter the RMC is only outlined roughly. Further details regarding the parts and the application scenario can be found in [12].

5.5.1 Usage Scenario

Aim of the RMC is to structure the multitude of models and standards shown in Section 5.3 and to construct a suitable frame of reference. The Energy RMC should provide software product managers with a structured access via a repository (knowledge base) and support the administration. The reference model catalog should enable software product managers to identify relevant information sources for their specific models and to improve them individually. On the basis of linking their own models to the reference model catalog, functional coverage analyses can be carried out and ideas for supporting functionality can be gained. In addition, changes within the sources which affect own models can be identified. In a potential usage scenario, various stakeholders participate in the construction, maintenance, and usage of the Energy RMC. These stakeholders may belong to a single organization or to several.

5.5.2 Components

The Energy RMC basically consists of four components, see Figure 5.7. The three most important parts of the catalog are the functional reference model *(FRM)*, *sources*, and *classification* criteria. These correspond to the structure, main and access parts of the reference model catalog concept by [10]. For the structure part a FRM is used.

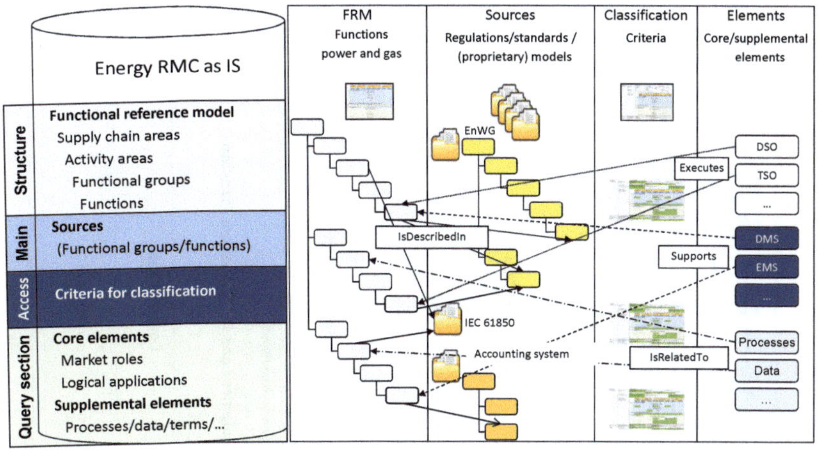

Fig. 5.7 Overview of the Energy RMC concept

A FRM is defined, according to [20], as a list of functions which describe a functional application area (in this case tasks of enterprises in the energy sector). The concept of [10] is extended by a query part consisting of core and additional elements. A short explanation of the single parts will follow. A more detailed description including the meta model can be found in [12]. As the modeling of functions is an important aspect in the development of application systems [19], functions also play a central role in the Energy RMC. The FRM is the core part of the catalog containing supply chain areas, activity areas, function groups, and functions (Figure 5.7).

The FRM describes specific supply chain areas and activities from the point of view of utility companies. Figure 5.8 shows the specific matrix of the FRM consisting of supply chain areas (vertical) and activities (horizontal). The FRM has to be seen as a functional hierarchy like the table of contents of a book, containing links (here relations) to the additional components (Figure 5.7).

Information sources (main part) and their classification (access part) are an essential aspect of the Energy RMC. Contrary to [10], not only reference models but also other information sources (Section 5.3) are classified. Sources are linked with the FRM (Figure 5.7 relation “IsDescribedIn”, arrows from left to right) and if necessary described within the catalog as a hierarchical structure consisting of function

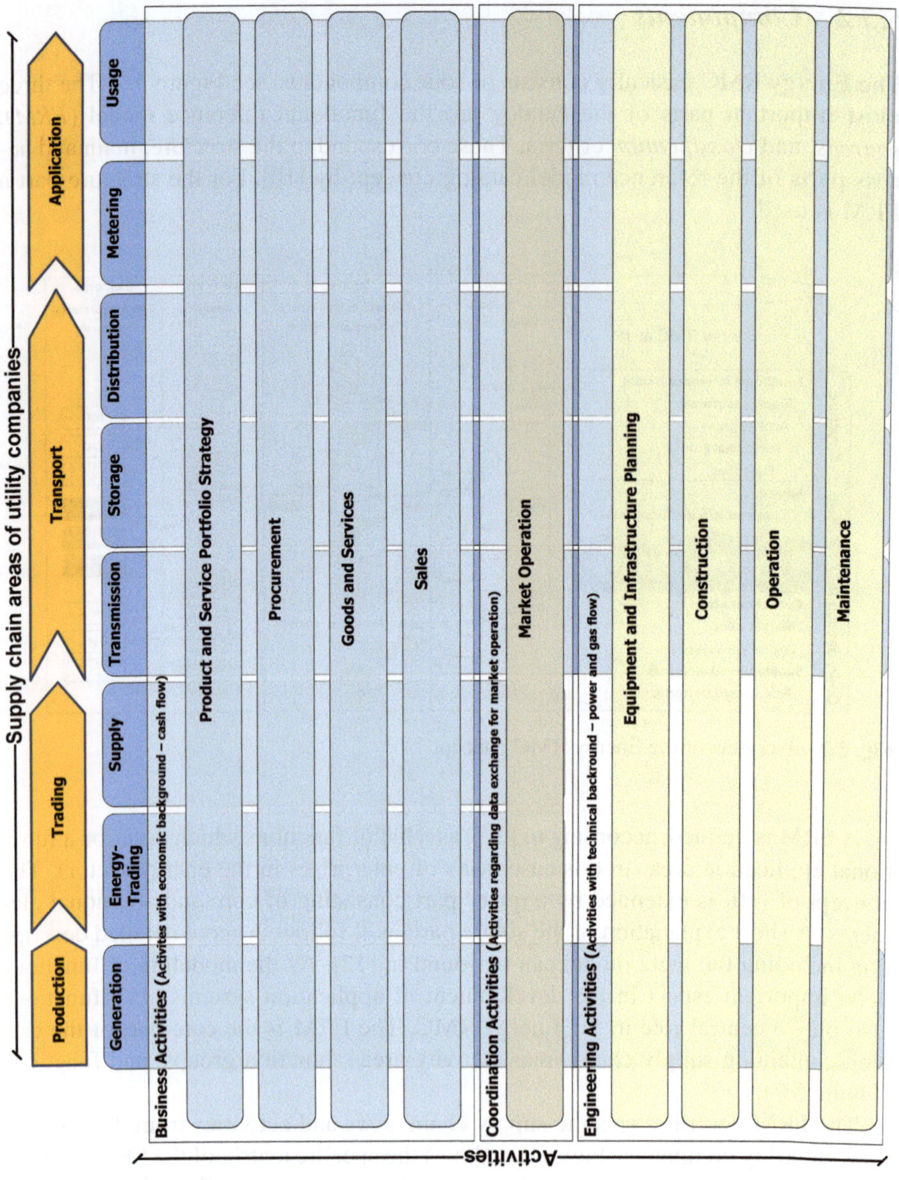

Fig. 5.8 The functional matrix of the FRM

groups and functions, where function groups may again contain function groups themselves.

The detailed description of sources as a hierarchy is used to identify relevant sources by means of the FRM or to compare the functional coverage of two sources. The suitable degree of details is chosen under economic points of view according to the perceived relevance of the source for the catalog users. For the EnWG, the central law for the German energy market, Figure 5.7 for example shows a functional hierarchy linked to the FRM on the lowest level, whereas for the IEC 61850 (automation standard) only a link on a higher level exists.

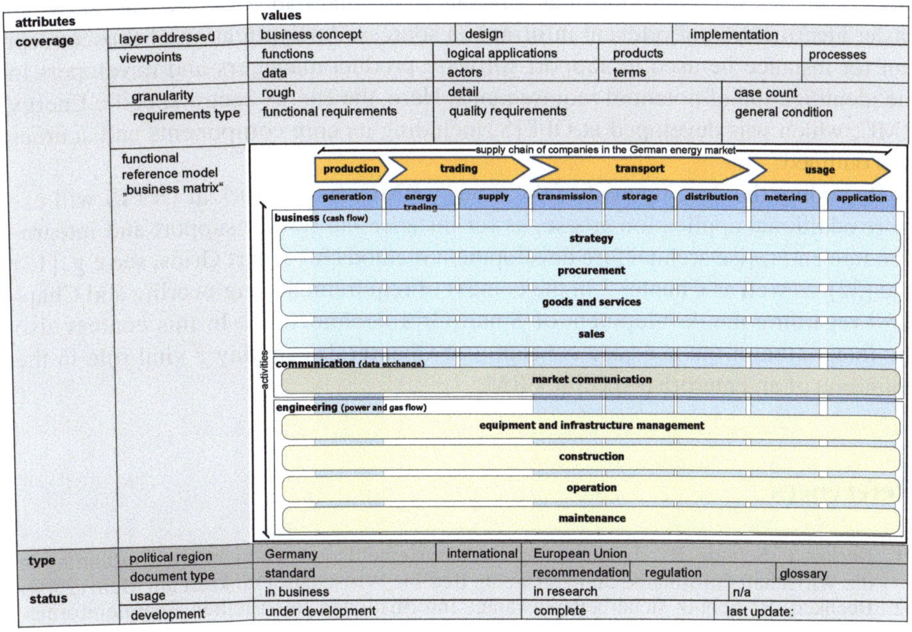

Fig. 5.9 Morphological box of the classification criteria

The classification of sources is based on established criteria for reference models and standards, see [25], [33], [35], [3], and [11]. Figure 5.9 shows the three essential attributes for the classification and possible effects: coverage (covered areas and views), type (origin and the form of the source), and status (usage and status of development). Figure 5.7 indicates that a classification is carried out for each source. The query part consists of core and additional components. Market roles (such as distribution system operator (DSO)) and logical applications (like distribution management systems (DMS)) as core elements are in the focus of the RMC and are therefore covered as completely as possible. In contrast, additional elements (products, processes, business objects, and definitions) enriching the RMC and functioning as optional elements are described only when needed (typically with name and description). These often only roughly described additional elements also enable quality assurance to ensure that of all relevant functions are covered. In the same way that sources are linked depending on their relevance, core elements (relation "Executes" and "Supports") are positioned on the lowest level of FRM, and additional elements (relation "IsRelatedTo") on a higher level.

5.6 Summary and Outlook

In this chapter the need for using existing information models to speed up requirement analysis in the energy sector was motivated. Apart from introducing information models and related concepts for structuring information models, reference model catalogs were introduced as a means to maintain an overview and allow an easier identification of relevant information sources. An application of this concept can for instance be used to support software product managers and developers in the identification of potential requirements. Here, the energy sector specific Energy RMC, which was developed at OFFIS, including its core components and sources was outlined.

Regarding the development of the Energy RMC, future work at OFFIS will explore additional application arease, as for instance the further support and integration into enterprise architecture development methods for Smart Grids, see e.g. [13] and [32], as well as Chapter 2 in the context of requirements engineering and Chapter 4 regarding the development of Smart Grid Architectures. In this context also the integration of the use case concept (see Chapter 3) can play a vital role in the definition of an enterprise-specific FRM.

References

1. Becker, J., Schütte, R.: Handelsinformationssysteme: Domänenorientierte Einführung in die Wirtschaftsinformatik, 2nd vollst edn. Redline Wirtschaft, Frankfurt am Main (2004)
2. Beenken, P.: Schutz sicherheitsrelevanter Informationen in verteilten Energieinformationssystemen. Dissertation, Carl von Ossietzky Universität Oldenburg (2011)
3. Braun, R., Esswein, W.: Classification of Reference Models. In: Decker, R., Lenz, H.-J. (eds.) Advances in Data Analysis Proceedings of the 30th Annual Conference of the Gesellschaft für Klassifikation e.V, March 8-10 (2006); Studies in Classification, Data Analysis, and Knowledge Organization, pp. 401–408. Springer (2007)
4. Brinker, W.: The Changing Structure of the Utility Industry from Its own Perspective. In: Bausch, A., Schwenker, B. (eds.) Handbook Utility Management, pp. 207–222. Springer (2009)
5. vom Brocke, J.: Design Principles for Reference Modeling. Reusing Information Models by Means of Aggregation, Specialisation, Instantiation, and Analogy. Reference Modeling for Business Process Analysis, 47–75 (2007)
6. Bundesministerium für Wirtschaft und Technologie: Endenergieverbrauch nach Energieträgern: Deutschland: Quelle: Arbeitsgemeinschaft Energiebilanzen, Stand: August 2008; letzte Änderung: October 08, 2008 (2008)
7. Bundestag, D.: Gesetz über die Elektrizitäts- und Gasversorgung (Energiewirtschaftsgesetz - EnWG) (2012)
8. ENTSO-E, European Federation of Energy Traders (EFET), energy Business Information eXchange (ebIX): The Harmonised Elecricity Market Role Model. Tech. rep. (2011)
9. Fettke, P.: Referenzmodellevaluation: Konzeption der strukturalistischen Referenzmodellierung und Entfaltung ontologischer Gütekriterien. In: Wirtschaftsinformatik - Theorie und Anwendung, vol. 5. Logos-Verl., Berlin (2006)
10. Fettke, P., Loos, P.: Der Referenzmodellkatalog als Instrument des Wissensmanagements - Methodik und Anwendung. In: Becker, J., Knackstedt, R. (eds.) Wissensmanagement mit Referenzmodellen. Konzepte für die Anwendungssystem- und Organisationsgestaltung, pp. 3–24. Springer, Berlin (2002)

11. Fettke, P., Loos, P.: Classification of reference models - a methodology and its application. Information Systems and e-Business Management 1(1), 35–53 (2003)
12. González, J.M.: Ein Referenzmodellkatalog für die Energiewirtschaft. Ph.D. thesis, Universität Oldenburg (2012)
13. González, J.M., Dänekas, C., Trefke, J., Uslar, M.: Supporting Interoperability in Smart Grids. In: I-ESA 2012 - Interoperability for Enterprise Systems and Applications (6th International Conference) (2012)
14. Haas, R., Redl, C., Auer, H.: The Changing Structure of the Electric Utility Industry in Europe: Liberalisation, New Demands and Remaining Barriers. In: Bausch, A., Schwenker, B. (eds.) Handbook Utility Management, pp. 169–192. Springer (2009)
15. Karagiannis, D., Kühn, H.: Metamodelling Platforms. In: Bauknecht, K., Tjoa, A.M., Quirchmayr, G. (eds.) EC-Web 2002. LNCS, vol. 2455, p. 182. Springer, Heidelberg (2002)
16. Koch, M., Baier, D.: Handel im liberalisieren Strommarkt, pp. 43–62 (2003)
17. Kurth, M.: The Changing Structure of the Utility Industry from the Perspective of Regulation Authorities. In: Bausch, A., Schwenker, B. (eds.) Handbook Utility Management, pp. 193–206. Springer, Berlin (2009)
18. Miksch, K.: Geschäftsbereich Energieversorgungsunternehmen IDS Scheer (2004)
19. Myrach, T.: Funktionsmodellierung. In: Kurbel, K., Becker, J., Gronau, N., Sinz, E., Suhl, L. (eds.) Enzyklopädie der Wirtschaftsinformatik, Oldenbourg (2009)
20. Närman, P., Gammelgard, M., Nordström, L.: Functional Reference Model For Asset Management Applications Based on IEC 61968-1. In: Nordic Distribution and Asset Management Conference (2006)
21. Object Management Group: OMG Unified Modeling Language (OMG UML), Infrastructure. Tech. Rep. (August 2011)
22. OFFIS, SCC Consulting, management coaching, M.: Untersuchung des Normungsumfeldes zum BMWi-Förderschwerpunkt 'E-Energy - IKT-basiertes Energiesystem der Zukunft' (2009)
23. SAP: Industry-Specific SAP Business Maps: UTILITIES: Utilities-Edition 2008: SAP Solution Map (2008)
24. Scheer, A.W.: {B}usiness process engineering: {R}eference {M}odels for {I}ndustrial {E}nterprises, 2nd edn. Springer, Berlin (1994)
25. Schütte, R.: Grundsätze ordnungsmä{ß}iger Referenzmodellierung: Konstruktion konfigurations- und anpassungsorientierter Modelle, vol. 233. Gabler, Wiesbaden (1998)
26. Stachowiak, H.: Allgemeine Modelltheorie. Springer, Wien (1973)
27. Steinmüller, W.: Informationstechnologie und Gesellschaft: Einführung in die Angewandte Informatik. Wiss. Buchges. Darmstadt (1993)
28. Thomas, O.: Das Modellverständnis in der Wirtschaftsinformatik: Historie, Lieteraturanalyse und Begriffsexplikation: Heft 184
29. Thomas, O.: Das Referenzmodellverständnis in der Wirtschaftsinformatik: Historie, Literaturanalyse und Begriffsexplikation: Heft 187. Saarbrücken (2006)
30. Thomas, O.: Management von Referenzmodellen: Entwurf und Realisierung eines Informationssystems zur Entwicklung und Anwendung von Referenzmodellen, Logos-Verl., Berlin (2006)
31. Thomas, O.: Understanding the Term Reference Model in Information Systems Research: History, Literature Analysis and Explanation. In: Bussler, C.J., Haller, A. (eds.) BPM 2005. LNCS, vol. 3812, pp. 484–496. Springer, Heidelberg (2006)
32. Trefke, J., Dänekas, C., Rohjans, S., González, J.M.: Adaptive Architecture Development for Smart Grids Based on Integrated Building Blocks. In: 3rd IEEE PES Innovative Smart Grid Technologies (ISGT) Europe Conference (2012)
33. Vom Brocke, J.: Referenzmodellierung: Gestaltung und Verteilung von Konstruktionsprozessen. Advances in Information Systems and Management Science, vol. 4. Jan vom Brocke, Berlin (2003)

34. Vom Brocke, J., Fettke, P.: Referenzmodellierung. In: Kurbel, K., Becker, J., Gronau, N., Sinz, E., Suhl, L. (eds.) Enzyklopädie der Wirtschaftsinformatik, Oldenbourg, München (2009)
35. de Vries, H.J.: IT Standards Typology. In: Jakobs, K. (ed.) Advanced Topics in Information Technology Standards and Standardization Research, pp. 1–26. Idea Group Pub. (2006)
36. Zachman, J.A.: A framework for information systems architecture. IBM Systems Journal 38(2-3), 454–470 (1999)

Part III
Standards and Applications

Chapter 6
ICT and Energy Supply: IEC 61970/61968 Common Information Model

Michael Specht and Sebastian Rohjans

Abstract. This chapter deals with one of the most recommended ICT-standards for the power domain. In particular, the CIM and the appropriate standard series IEC 61970, IEC 61968, and IEC 62325 are introduced. The CIM is a powerful overall integration framework, which is historically grown and continuously improved in order to meet the latest requirements. Due to the fact that the CIM is designed as an abstract and generic model, the developments focus on two focal areas: On the one hand, a comprehensive data model is developed and maintained in UML. On the other hand, technology mappings are specified in order to make the overall model applicable. In general, the CIM is used for two major use cases: Message exchange based on XML serializations and exchange of power grid topologies serialized with RDF. Within this chapter all these aspects are covered and analyzed. Finally, the chapter concludes with a summary and an outlook.

6.1 Introduction and History

The Common Information Model (CIM) was originally developed by the Electric Power and Research Institute (EPRI) in the midst of the 90's. In the Control Center API (CCAPI) project, the CIM was designed to solve the problem of vendor lock-ins. For that reason, it offered an internal database model for Energy Management Systems (EMS) and Supervisory Control and Data Acquisition (SCADA) systems. Over the years, the CIM has outgrown its original purpose and now contains a pretty large domain ontology, which covers most topics in the power domain. Additionally, it serves as an integration model and delivers interface specifications as well as data serializations [1].

Michael Specht · Sebastian Rohjans
OFFIS – Institute for Information Technology, Escherweg 2, 26121 Oldenburg, Germany
e-mail: {michael.specht, sebastian.rohjans}@offis.de

M. Uslar et al.: *Standardization in Smart Grids*, POWSYS, pp. 99–114.
DOI: 10.1007/978-3-642-34916-4_6

The development of the CIM is currently carried out by the International Electrotechnical Commission (IEC) in the Technical Committee (TC) 57. The following Working Groups (WG) are concerned with the development of the CIM:

- WG 13 – “Energy Management System Application Program Interface (EMS–API)”
- WG 14 – “System Interfaces for Distribution Management (SIDM)”
- WG 16 – “Deregulated Energy Market Communications”
- WG 19 – “Long Term Interoperability within IEC TC 57 Working Groups”

Whereas the IEC is focusing on standardization tasks, another forum was founded to support utilities, vendors, or consultants in applying the CIM. This group is named CIM users group[1] (CIMug) and is the central point of information about the CIM. The CIMug provides recent news on the CIM and the latest versions of the electronic data model as well as it hosts meetings.

The EPRI takes on a continued role in the CIM environment and conducts research in areas where the CIM needs additional definitions or visibility. Furthermore, the EPRI coordinates the annual interoperability testing (see Section 9.3 for more information). These formal test methods address the interoperability of EMS and third-party vendor products. Interoperability testing establishes that products from different participant vendors can exchange information based on the use of CIM standards. The CIM standard family is divided into the following different series:

- **IEC 61970** “Energy Management System Application Program Interfaces (EMS-API)” [3]

  The extensive, basic data model defined within IEC 61970-301 represents the main part of this standard series and includes most of the objects required to model power networks. Additionally, the IEC 61970 contains the Component Interface Specifications (CIS)—defining how the platform independent data models and the generic interfaces can be used in combination with communication standards—as well as the Generic Interface Definitions (GID)—focusing on the status of exchanged data and its use compliant to CIM semantics—. One of the main objectives of the CIM is to provide a platform independent data model. In order to make this model applicable, mappings to specific technologies like Resource Description Framework (RDF), Extensible Markup Language (XML), and Web Ontology Language (OWL) are specified.
- **IEC 61968** “Application Integration at Electric Utilities – System Interfaces for Distribution Management” [7]

  In contrast to the IEC 61970, the IEC 61968 focuses more on virtual objects required for business use cases like billing, markets, or network extension planning. A main part of this series is the Interface Reference Model (IRM), which specifies use cases, interfaces, and messages in the power domain. The base data model, defined in IEC 61970-301, is extended by further objects specified within IEC 61968-11.

[1] `http://cimug.ucaiug.org`

- **IEC 62325** "Framework for energy market communications" [5]
  IEC 62325 represents a set of standards describing a framework for energy market communications. Its main parts are covering the communications between market participants and market operators. Additionally, two market styles are supported: "European-style markets" and "US-style markets". Due to the importance of market communication within future power systems, this topic is considered separately in Section 13.2.6.

Figure 6.1 provides an overview on a classification of the single parts and subparts of the introduced standard series. The remainder of this chapter is oriented towards this classification and each building block is described more detailed. However, the topmost building block is an exception since it includes the general parts of the standard series dealing with glossaries and basic principles, which have already been introduced.

6.2 Data Models

One of the basic parts of the CIM is the data model, which can be seen as a domain ontology for the power domain. Basically, the data model is capable of converting real power domain objects into a Unified Modeling Language (UML) data model as illustrated in Figure 6.2.

The data model is the lowermost and basic building block of those shown in Figure 6.1. The overall data model itself is split into three subparts, which are IEC 61970-301–"CIM Base", IEC 61968-11–"Distribution Information Exchange Model", and IEC 62325-301–"Data Model for Market Extension". Each part has different UML packages with different objects and focuses.

Table A.3 in Chapter A provides an overview on the IEC 61970-301 data model packages and on the content of each package as well. This data model is the crystallization point of the CIM and includes the basic components to represent power domain specific objects.

The IEC 61968-11 data model packages are addressed in Chapter A (see Table A.1). The IEC 61968 focuses on business cases. For this reason, the data model is oriented towards business case-related objects.

The packages specified in IEC 62325-301 are dealt with in Table A.2 in Chapter A. This packages mainly concentrate on objects required to model market communication messages.

The different UML data models are maintained in different WGs and are merged regularly in order to annually provide with a complete, joined CIM data model release[2]. This model can be obtained at the CIMug website. The modeling platform

---

[2] The latest version "iec61970cim15v33_iec61968cim11v13_iec62325cim01v07" has been lastly modified in April 2012. The version's name consists of information about the major and minor releases of the single parts, e.g., for the IEC 61970 part it is the 15th major release with its 33rd minor release.

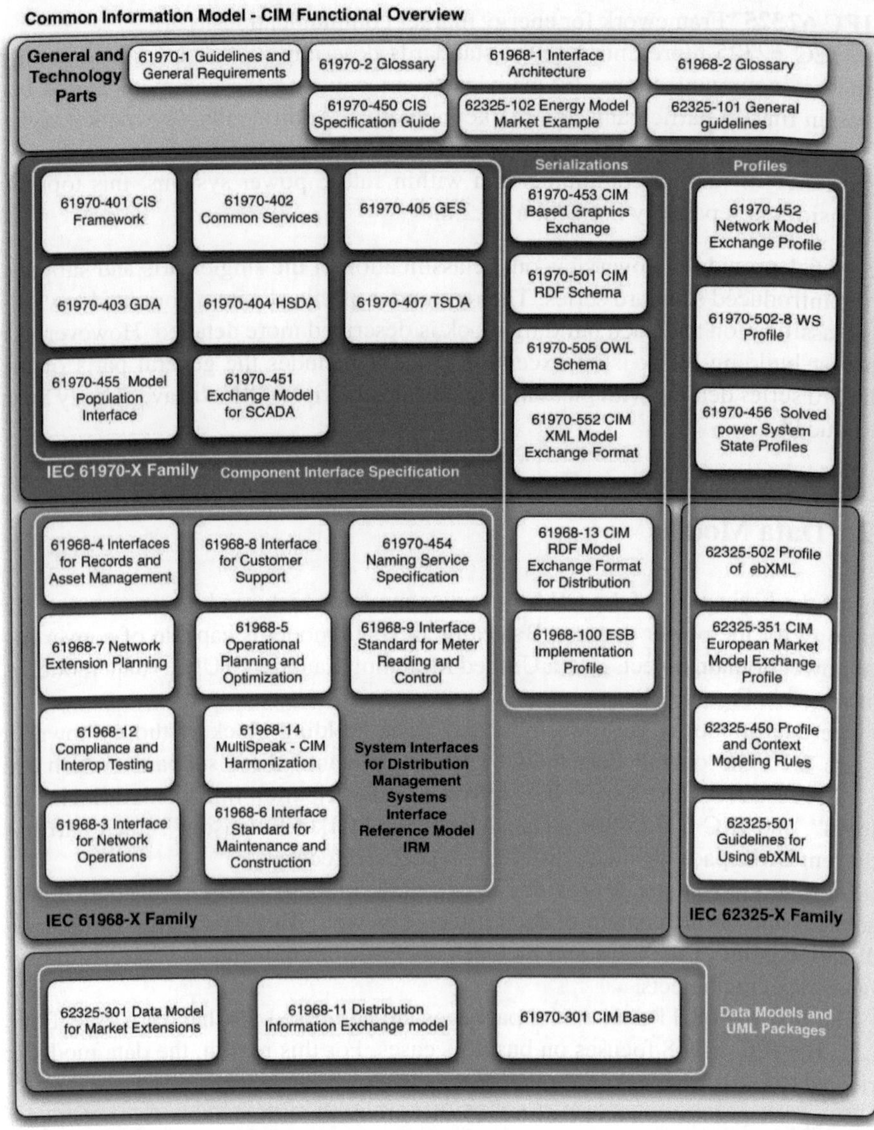

Fig. 6.1 Overview on existing CIM standards [21]

Enterprise Architect (EA) from Sparx Systems[3] has been chosen to maintain the CIM UML model. Section 6.7 provides more detailed information about this tool.

In total, the overall UML model was counting over 900 classes in the previous version 13 [20] and includes about 1300 classes in the current version 15. This emphasizes the continuously increasing size of the data model, which is explained

[3] www.sparxsystems.com.au/products/ea/index.html

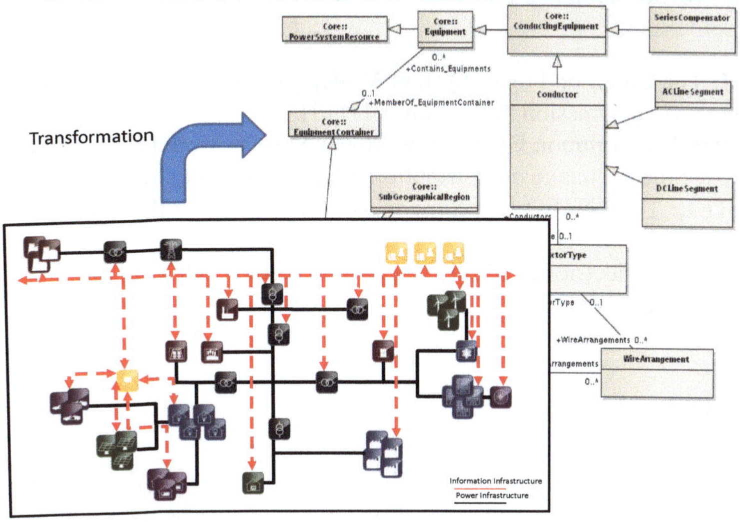

Fig. 6.2 Transformation from real world objects into the CIM data model

by new arising requirements postulated by the dynamically extending scope of the power domain.

All classes and associations have additional descriptions realized as annotated meta data, which aim at helping users to identify desired objects. The classes' attributes contain such descriptions as well. This additionally fosters the handling of the data model.

6.3 Profiles

The next building block to be analyzed includes CIM profiles. As mentioned in Section 6.2, the basis of the CIM is a comprehensive data model. Although the extensive size of the data model providing with almost all important objects can be seen as an advantage, this size complicates the data model's application. In order to solve this problem, it is possible to define and generate profiles for specific scenarios. A CIM profile is a real subset of the original data model. It includes classes and associations required for the scenario. Furthermore, attributes of the considered classes can be defined as optional or mandatory. Finally, additional restrictions such as cardinalities of associations can be specified.

Creating profiles is an established and recommended means to cope with the CIM data model. Profiles can be created individually by single users in enterprise-specific contexts or officially by working groups or other organizations. Official profiles address a broader audience and thus have a higher influence. The most relevant official profiles are:

- **CPSM:** The Common Power System Model (CPSM) is used in the USA for the exchange of transmission system models [11].
- **CDPSM:** The Common Distribution Power System Model (CDPSM) is used in Europe for the exchange of distribution power system models [8, 15].
- **ENTSO-E:** The ENTSO-E[4] profile is used in Europe for the exchange of transmission system models.
- **ERCOT:** The Electric Reliability Council of Texas[5] (ERCOT) profile is an intra-corporate data model.

6.4 Serializations

The building block concerned with serializations of the abstract data model is driven by application cases for the CIM. In particular, two major use cases are defined for the CIM [18]. The first use case deals with CIM-based XML message exchange and the second use case addresses the exchange of power grid topology data in RDF format. However, to make the CIM applicable, mappings have to be defined and standardized. The developed mappings are tailored for different use cases. In fact, the following mappings are considered:

- **Java Messaging System (JMS) and Enterprise Service Bus (ESB):** IEC 61968-100 [13] specifies how standardized message payloads can be utilized based on Web Services, JMS, and ESB technologies.
- **MultiSpeak (V 4.1):** A harmonization of CIM and MultiSpeak (V 4.1) is in the scope of IEC 61968-14 [14]. MultiSpeak is a standard for software interoperability in the power domain mainly applied in the US. The mapping is performed on profile level. This means that each CIM profile is mapped onto an appropriate MultiSpeak element. Thus, interfaces can be translated and gaps on one of the sides can be identified.
- **XML/RDF:** IEC 61970-501 [6] specifies how the UML data model can be serialized as machine-readable XML representation. It defines formats and rules in order to enable generation of RDF compliant documents. The considered solution is both, machine-readable and human-readable. Moreover, it can be accessed using any tool that supports the Document Object Model (DOM) API, is self-describing, and takes advantage of Web standards provided by the World Wide Web Consortium (W3C).
- **CIM/XML:** How the CIM RDF schema could be used to exchange power system models is described in IEC 61970-552 [12]. The standard defines a CIM/XML model exchange format.

[4] `http://www.entso-e.org/`

[5] `http://www.ercot.com/`

- **OPC Unified Architecture (OPC UA):** An integration of CIM and OPC UA is under development. In future, IEC 61970-502-8 [9] will cover the mapping of CIM objects onto the OPC UA information model. The mapping aims at applying CIM semantics within a server-client Service-Oriented Architecture (SOA).
- **Web Ontology Language (OWL):** The scope of future IEC 61970-505 is supposed to include a description of an OWL schema for CIM[6].
- **Electronic Business using XML (ebXML):** General guidelines related to the integration of ebXML technologies and architectures are described in IEC 62325-501 [2]. Therefore, migration scenarios and implementation examples are provided.

6.4.1 CIM-Based XML Message Exchange

The XML-based message exchange is mainly described in the IEC 61968 standard series. Message exchange based on CIM semantics can either be point-to-point or via ESB systems. Due to the fact that using CIM semantics in combination with ESB systems is the more commonly used alternative, this option will be focused on.

The message exchange itself can be vary in complexity. Possible applications range from simple request/reply messages to nested chain message exchange with many asynchronous replies or event messages. Regardless of the complexity of the message exchange, the basic message structure is always the same and is standardized in IEC 61968-1 [7]. It is recommended by the IEC 61968-1 standard to use the following elements in order to clearly identify a message and the appropriate recipient:

- **Verb:** Identification of the type of action; limited enumeration strings like create, delete, etc.
- **Noun:** Identification of the type of the payload
- **Payload:** Containing the relevant data regarding the information exchange

The message structure itself is formalized as XML Schema and is depicted in Figure 6.3. The *header* is mandatory for all messages (except for fault response messages) and uses a common structure for all service interfaces. The optional *request* parts define commonly used parameters, which are required to qualify requests and to identify specific objects for actions like delete or cancel. *Reply* is only required for response messages to indicate details on the success, failure or error. The *payload* is often required and is used to convey the message information as a consequence of the *verb* and *noun* from the message header. All introduced elements include further objects and the appropriate formalized XML Schema can be found in [7].

Due to the standardized message structure, the focus for developing messages is on the message's payload. The IEC 61968-9 "Application integration at electric utilities – System interfaces for distribution management – Part 9: Interface for

[6] As presented in February 2011 at the WG 13 meeting in Santa Clara.

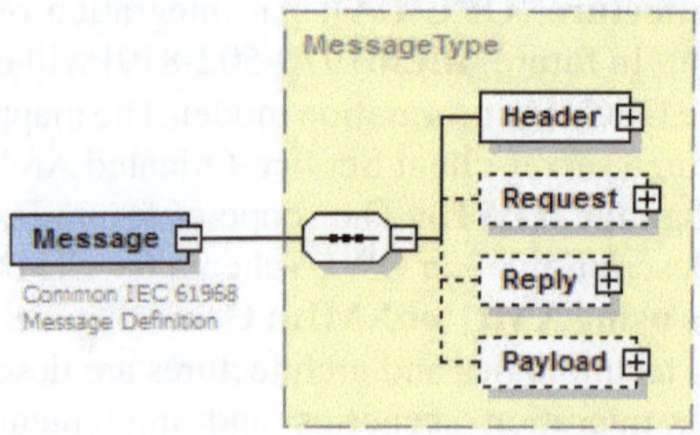

Fig. 6.3 XML Schema of the basic CIM message structure from [13]

meter reading and control" standard [10] is one of the few standards, which specify standardized payloads for the CIM. Figure 6.4 shows the graphical representation of the *EndDeviceEvent* message's payload as XML Schema, which serves as an example in this chapter. The *EndDeviceEvent* message is primarily used to convey events such as device health events, power quality events, and outage events at end-devices (e.g., smart meters).

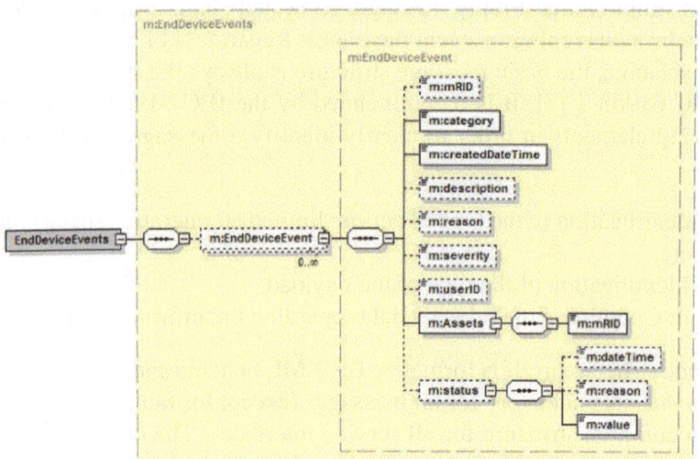

Fig. 6.4 Graphical representation of the *EndDeviceEvent* message as XML Schema [10]

As depicted in Figure 6.4, only *category*, the timestamp for *createdDateTime*, and the *mRID* of the *assets* are mandatory objects. Furthermore, *value* is mandatory if the optional *status* exists. A message payload based on this schema implemented with the at least required exemplary data is shown in Listing 6.1.

Listing 6.1 *EndDeviceEvent* message payload

```
<m:EndDeviceEvents xsi:schemaLocation=
"http://iec.ch/TC57/2009/EndDeviceEvents#_schema.xsd"
xmlns:m="http://iec.ch/TC57/2009/EndDeviceEvents#"
xmlns:xsi="http://www.w3.org/2001/XMLSchema-instance">
 <m:EndDeviceEvent>
   <m:category>Outage Alarm</m:category>
   <m:createdDateTime>2012-04-30T21:30:47.0Z</m:createdDateTime>
   <m:Assets>
     <m:mRID>42_1337</m:mRID>
   </m:Assets>
 </m:EndDeviceEvent>
</m:EndDeviceEvents>
```

This payload describes an outage alarm event, which is detected on 2012-04-30 at 21:30:47 on the device with the *mRID* "42_1337". In order to send this information, it is necessary to embed this payload into the before mentioned message structure. A possible, complete message is shown in Listing 6.2. This message can be received and processed by a properly configured ESB.

Listing 6.2 *EndDeviceEvent* instance message

```
<?xml version="1.0" encoding="utf-8"?>
<soap:Envelope xmlns:soap="schemas.xmlsoap.org/soap/envelope/"
xmlns:xsi="http://www.w3.org/2001/XMLSchema-instance"
xmlns:xsd="http://www.w3.org/2001/XMLSchema">
  <soap:Header>
    <Password>User1</Password>
    <Login>User1</Login>
  </soap:Header>
  <soap:Body>
    <MessageHeader>
      <Verb>created</Verb>
      <Noun>EndDeviceEvent</Noun>
    </MessageHeader>
    <MessagePayload>
      <m:EndDeviceEvents xsi:schemaLocation=
        "http://iec.ch/TC57/2009/EndDeviceEvents#_schema.xsd"
      xmlns:m="http://iec.ch/TC57/2009/EndDeviceEvents#"
      xmlns:xsi="http://www.w3.org/2001/XMLSchema-instance">
        <m:EndDeviceEvent>
          <m:category>Outage Alarm</m:category>
          <m:createdDateTime>2012-04-30T21:30:47.0Z
            </m:createdDateTime>
          <m:Assets>
            <m:mRID>42_1337</m:mRID>
          </m:Assets>
        </m:EndDeviceEvent>
      </m:EndDeviceEvents>
    </MessagePayload>
  </soap:Body>
</soap:Envelope>
```

6.4.2 Topology Data Exchange

This use case deals with the important task to exchange power grid topology data between different parties. In todays intermeshed electricity networks—managed by different network operators—it is inevitable to exchange data about the networks to not threaten the quality of service and to prevent disturbances like blackouts.

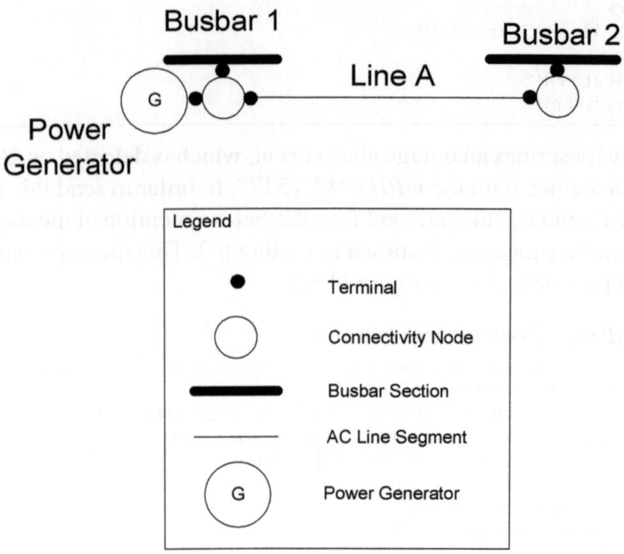

Fig. 6.5 Single line diagram describing an exemplary topology including CIM objects

To work towards this goal, a standardized exchange of power grid topology data is enabled through the CIM. It offers a serialization from objects (both, physical and virtual objects) in the XML-based data format RDF. Each object from the UML data model can be transformed into an equivalent RDF object.

Figure 6.5 shows an example for a small-sized power grid in single line diagram format enriched with special CIM objects. These special objects are used to connect physical objects with *terminals* and *connectivity nodes* (virtual objects). This is due to the CIM serialization, which does not allow to directly connect physical objects. As the legend of this figure states, five different objects are included in the exemplary power grid. On the left-hand side, a *power generator* and a *bus bar section* are located at the same place. These objects are connected via *terminals* and *connectivity nodes*. Another *bus bar section* is located on the right-hand side. Both *bus bar sections* and the appropriate *terminals* and *connectivity nodes* are connected via an *AC line segment*.

The objects required to serialize this small-sized power grid are shown in Figure 6.6 using UML class diagram notation. Most of the classes are self-explanatory due to their names. Only the *power generator* as part of the single line diagram is represented by the class *EnergySource*, which is a rather abstract class.

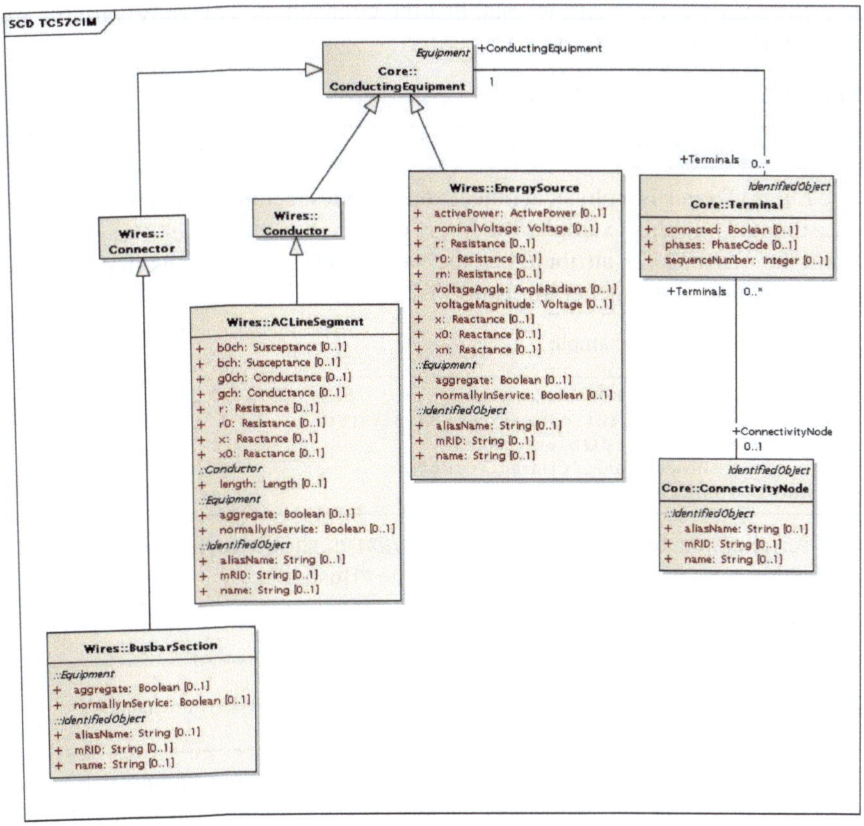

Fig. 6.6 Used CIM elements from the UML data model

As shown in Figure 6.6, all UML objects in this example are interconnected. This structure allows the creation of a graph-like topology in form of RDF. Each object has to be serialized individually and references to other objects have to be considered as well. The attributes and associations are represented using dot notation in addition to the class name. Each object will be briefly introduced:

The minimal *Terminal* object (see Listing 6.3) consists of a RDF ID ("X") and references to elements, which it connects. These are, on the one side, a *ConductingEquipment* (e.g., bus bar) with a reference named "XA" and on the other side, the CIM-specific *ConnectivityNode* with a reference named "XB".

Listing 6.3 Minimal *Terminal* example

```
<cim:Terminal rdf:ID="X">
  <cim:Terminal.ConductingEquipment rdf:resource="#XA" />
  <cim:Terminal.ConnectivityNode rdf:resource="#XB" />
</cim:Terminal>
```

In contrast to the *Terminal*, the *ConnectivityNode* used in this example has no additional attributes. Due to the graph structure, the reference from *Terminal* to

ConnectivityNode is sufficient to establish the connection. The only variable used is the RDF ID (see Listing 6.4).

Listing 6.4 *ConnectivityNode* example

```
<cim:ConnectivityNode rdf:ID="Y"></cim:ConnectivityNode>
```

The *EnergySource* is a physical object and includes technical data for a generator (see Listing 6.5). In this example, the *EnergySource* has the name “Power Generator” and the *activePower* attribute has the value “400”. This means that the active power production of this generator is 400 W.

Listing 6.5 *EnergySource* example

```
<cim:EnergySource rdf:ID="A_G">
    <cim:IdentifiedObject.name>Power Generator
        </cim:IdentifiedObject.name>
    <cim:activePower>400</cim:activePower>
</cim:EnergySource>
```

The serialization for the class *BusBarSection* is shown in Listing 6.6. In this example, this class is only specified by its name “Busbar”.

Listing 6.6 *BusbarSection* example

```
<cim:BusbarSection rdf:ID="A_B1">
  <cim:IdentifiedObject.name>Busbar 1</cim:IdentifiedObject.name>
</cim:BusbarSection>
```

The last component is the *ACLineSegment*, which describes technical information such as resistance (r), reactance (x), susceptance, length, and name of a line (cable) (see Listing 6.7).

Listing 6.7 *ACLineSegment* example

```
<cim:ACLineSegment rdf:ID="A_A1">
  <cim:Conductor.length>2500</cim:Conductor.length>
  <cim:Conductor.r>0.3125</cim:Conductor.r>
  <cim:Conductor.x>0.28</cim:Conductor.x>
  <cim:Conductor.bch>235.45</cim:Conductor.bch>
  <cim:IdentifiedObject.name>Line A</cim:IdentifiedObject.name>
</cim:ACLineSegment>
```

Combining the information from Listings 6.3–6.7 results in a complete RDF serialization of the example illustrated in Figures 6.5 and 6.6. The whole serialization in RDF can be found in Annex B.

6.5 Component Interface Specifications (CIS)

The CIS-framework is a building block, which is basically specified in IEC 61970-401 [4]. The pursued goal of the CIS is to define interfaces for components and applications. In the scope of the CIS are those standards and applications, which intend to either exchange data among each other or to access publicly available data. The framework is divided into one part specifying generic services for data exchange and a second part defining information content.

The GID focuses on the status of the data and their use compliant to CIM semantics. The considered interfaces comprise Generic Data Access (GDA), High-Speed Data Access (HSDA), Generic Eventing and Subscription (GES), and Time Series Data Access (TSDA). The implemented common services are based on the GID and classified as resource identifier services (identifying classes, class attributes, and object instances in systems), resource description services (encoding values associated with classes, class attributes, and object instances), and view services (representing classes, class attributes, and object instances by hierarchies and tree structures).

6.6 Interface Reference Model (IRM)

The IRM building block is a vital part of the CIM and is basically standardized in IEC 61968-1 [7]. Beside the ESB or an alternative middleware, systems, which are part of a Distribution Management System (DMS), are classified by their functionalities. The interfaces, which connect the systems with the middleware layer, are dealt with in terms of their requirements. Within the further subparts of the standard series, these interfaces are covered and discussed technology-independent. In detail, the standards IEC 61968-3 to -10 describe the most important business processes from the IRM for message exchange covering the following functions:

- Monitoring and control of equipment for power delivery
- Management processes to ensure system reliability
- Voltage control
- Demand side management
- Outage management
- Work management
- Automated mapping
- Facilities management

6.7 Tooling

The CIM is directly modeled using the UML tool EA provided by Sparx Systems[7], which thus is one of the most influencing CIM-related tools. Besides, providing the capability to browse the UML model, it also offers the functionality to use specialized Add-Ins, which are mainly created by the CIM community. For example, CIMContextor[8] is an Add-In which helps to transform the UML model directly to

[7] `www.sparxsystems.de`

[8] `www.cimcontextor.net`

schemas and CIMEA[9] is an Add-In fostering the design and modeling of CIM-based artifacts such as RDF and XML Schema Definition (XSD) message generation.

Figure 6.7 shows a screenshot of a CIM model opened with the Sparx EA UML tool.

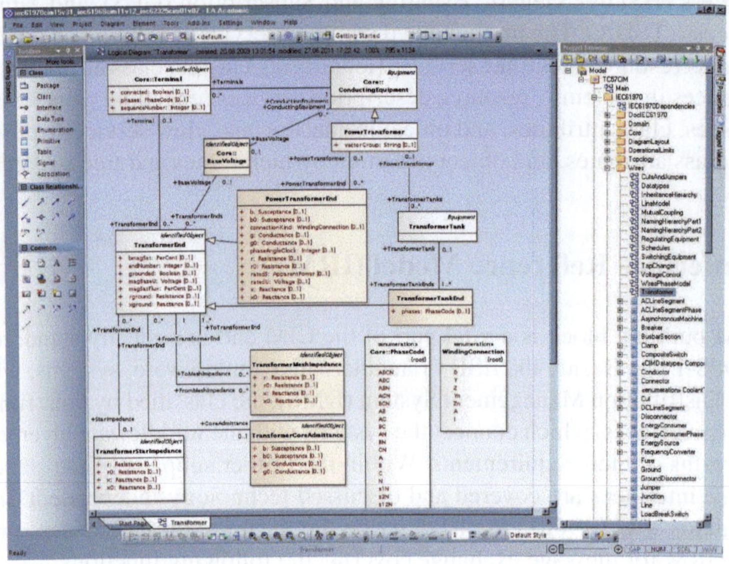

Fig. 6.7 CIM model in Sparx EA

Another relevant tool, which facilitates the application of the CIM, is CIMTool[10] . This is a freeware solution based on the Eclipse framework. It manages the creation of CIM profiles and message schema creation as well as validation. CIMTool is widely used in the CIM community and covers important tasks in the CIM domain like creating OWL ontologies.

CIMSpy is a commercial tool, which is either free of charge (limited features) or can be purchased to obtain all available features. The feature list includes the import of CIM/XML (RDF) topologies including their visualizations in single-line diagrams (see Figure 6.8), browsing the elements, make load-flow analysis, and the validation of CIM/XML (RDF) topologies in terms of both, syntax and in relation to existing profiles.

Finally, CIMbaT is an EA Add-On that can be used to create OPC UA Address Space [16]. In combination with SDKs provided by the OPC UA Foundation[11], OPC UA server applications being based on CIM semantics can be created. This tool is also described in detail in Chapter 12.

---

9 `www.cimea.org`

10 `www.cimtool.org`

11 `http://opcfoundation.org`

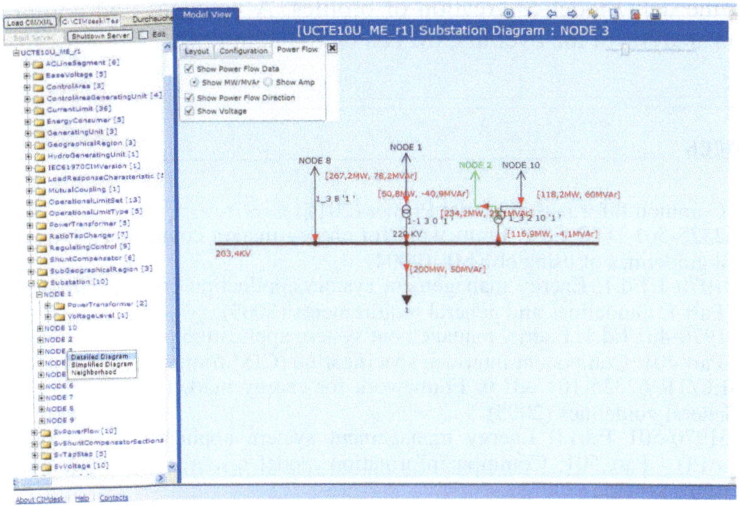

Fig. 6.8 CIMSpy single line diagram for load flow analysis

In general, the tool support for CIM is very dynamic. It changes and advances every year. In order to use the CIM with high efficacy, the latest tools and versions are recommended to be used.

6.8 Conclusion and Outlook

The CIM is one of the predominant and most recommended IT-standards on the application level in the power domain [19, 17]. Although the CIM was developed in the US, it gains more and more acceptance in Europe. Organizations like ENTSO-E (see Section 13.2.4) are creating CIM profiles and prescribe the profile for its members. Due to the active and continuously growing community, the importance of the CIM will still increase in the future. Future extensions include the creation of weather specific data models and extensions for Distributed Energy Resources (DER) and Plug-in Electric Vehicles (PEV). The development of interfaces to other systems like home automation will also be subject to future work.

Harmonizing the CIM with other existing or future power grid IT-standards is another focal topic. Especially WG 19 in the IEC TC 57 is concerned with solutions regarding harmonization with IEC 61850 and other standards. Furthermore, using CIM semantics for automation standards like OPC UA is frequently discussed and gains more and more momentum.

Concluding, the CIM already plays a key role in terms of IT-applications in power grids and it will strengthen its position in the future. It provides with all relevant capabilities to build the semantic basis for future power grids. Moreover, well-defined methodologies have been developed in order to make the CIM applicable, e.g., for extending the data model or creation of profiles. A practical and comprehensive application analysis of the overall CIM can be found in [21].

References

1. EPRI: Common Information Model Primer (2011)
2. IEC: 62325-501 DTR Ed.1: Framework for energy market communications - Part 501: General guidelines of using ebXML (2004)
3. IEC: 61970-1 Ed.1: Energy management system application program interface (EMS-API) - Part 1: Guidelines and general requirements (2005)
4. IEC: 61970-401 Ed.1: Energy management system application program interface (EMS-API) - Part 401: Component interface specification (CIS) framework (2005)
5. IEC: IEC/TR 62325-101 ed1.0: Framework for energy market communications - Part 101: General guidelines (2005)
6. IEC: 61970-501 Ed.1.0 Energy management system application program interface (EMS-API) - Part 501: Common information model resource description framework (CIM RDF) Schema (2006)
7. IEC: 61968-1: Application integration at electric utilities - System interfaces for distribution management - Part 1: Interface architecture and general requirements (2007)
8. IEC: 61968-13 Ed.1: Application integration at electric utilities - System interfaces for distribution management - Part 13: CIM RDF Model exchange format for distribution (2008)
9. IEC: 61970-502-8 Ed.1: Energy Management System Application Program Interface (EMS-API) - Part 502-8: CIM Data Services (Draft) (2008)
10. IEC: 61968-9 Ed.1: Application integration at electric utilities - System interfaces for distribution management - Part 9: Interface for meter reading and control (2009)
11. IEC: 61970-452 : Energy Management System Application Program Interface (EMS-API) - Part 452: CIM Transmission Network Model Exchange Profile Revision 6.7 (2009)
12. IEC: 61970-552-4: CIM XML Model Exchange Format (2009)
13. IEC: 61968-100 (Draft): Application integration at electric utilities - System interfaces for distribution management - Part 100: Implementation Profiles for IEC 61968 (2011)
14. IEC: 61968-14 (Draft): Application integration at electric utilities - System interfaces for distribution management - Part 14: MultiSpeak - CIM Harmonization (2011)
15. Lambert, E.: CDPSM: Common Distribution Power System Model: When, Why, What, How, Who? In: IEEE/PES Power Systems Conference and Exposition, PSCE (2011)
16. Rohjans, S., Piech, K., Uslar, M., Cabadi, J.F.: CIMbaT - Automated Generation of CIM-based OPC UA-Address Spaces. In: IEEE SmartGridComm. 2011, Brussels (2011)
17. Rohjans, S., Uslar, M., Bleiker, R., González, J.M., Specht, M., Suding, T., Weidelt, T.: Survey of Smart Grid Standardization Studies and Recommendations. In: First IEEE International Conference on Smart Grid Communications (2010)
18. Rohjans, S., Uslar, M., Piech, K., Cabadi, J.F., Santodomingo, R.: New Applications of the Common Information Model. In: International Symposium - The Electric Power System of the Future (2011)
19. Uslar, M., Rohjans, S., Bleiker, R., González, J.M., Suding, T., Specht, M., Weidelt, T.: Survey of Smart Grid Standardization Studies and Recommendations - Part 2. In: IEEE Innovative Smart Grid Technologies Europe (2010)
20. Uslar, M., Rohjans, S., Specht, M., González, J.M.: What is the CIM lacking? In: Innovative Smart Grid Technologies Conference Europe ISGT Europe 2010 IEEE PES (2010)
21. Uslar, M., Specht, M., Rohjans, S., Trefke, J., González, J.M.: The Common Information Model CIM: IEC 61968/61970 and 62325 - A Practical Introduction to the CIM. Springer (2012)

Chapter 7
Automation for the Smart Grid: IEC 61850 - Substation Automation and DER Communication

Mathias Uslar and Robert Bleiker

Abstract. Within this very chapter, we are going to provide a short overview on the large amount of applications and technologies involved with the IEC 61850 standard family. Starting with its origins, developed parts, and sub-parts, we will later focus on the information and communication side of this standard family and current developments in information modeling that lead to changes in how to apply the standard family. The focus of this chapter is the application of IEC 61850 in the scope of substation automation and Distributed Energy Resources and differences between those two major use cases.

7.1 Introduction to the IEC 61850 Standard Family

The aims of the development of the standard family "IEC 61850: Communication networks and systems in substations" are to achieve a better interoperability between Intelligent Electronic Devices (IEDs) in substations with special focus on multi-vendor systems, additionally increase the possibility for data exchange between sub-systems and to use this data to fulfill dedicated functions in the substation. The corresponding definition of interoperability is used analogously to the definition from the Institute of Electrical and Electronics Engineers (IEEE), which defines interoperability as the "ability of a system or a product to work with other systems or products without special effort on the part of the customer. Interoperability is made possible by the implementation of standards.". While interoperability can be seen as one of the basic goals of the standard family, a rather similar goal, achieving the so called interchangeability, is also desired. Interchangability can be defined as the possibility to replace one device from one vendor with a device from the very same

Mathias Uslar · Robert Bleiker
OFFIS – Institute for Information Technology, Escherweg 2, 26121 Oldenburg, Germany
e-mail: {uslar,bleiker}@offis.de

M. Uslar et al.: *Standardization in Smart Grids*, POWSYS, pp. 115–128.
DOI: 10.1007/978-3-642-34916-4_7

vendor or even by one from a different vendor without having to change configuration, functionality, or interfaces for the rest of the system. Interoperability may be seen as a pre-condition to achieve interchangeability. For this level of integration, two pre-conditions must be fulfilled. Apart from the so called syntactical interoperability, also the functional interoperability, i.e. standardizing the device's functionality would be required. Because this would typically interfere with the vendor's ability to bring competitive new devices to the markets, this integration level is out of scope for the IEC 61850.

7.2 History and Overview

The goal of achieving interoperability and the openness of the communication interface are sufficient for the IEC 61850 to be still a useful standard family. The openness of the communication, being mainly architectured through the use of the Abstract Communication System Interface (ACSI), makes for the future-proofness of the devices. Communication layers can be replaced with the latest state-of-the-art (MMS vs. Web Services) without having to change the internal device data model, improve the engineering efforts, and extend the original functionality (e.g. reporting) and data models of the device. The standard family's parts do not only focus on the communication as in a single layer of an ISO/OSI (International Organization for Standardization/Open Systems Interconnection) stack, but also address important system aspects like project management, domain-specific data modeling including extension rules, domain-specific services, a configuration language, and conformance tests. Figure 7.1 provides an overview on the various use cases of the different sub-standards of the IEC 61850 family in an abstract classification. Just like the CIM (Common Information Model), the IEC 61850 is not "one standard", but covers several important aspects and different communication mappings and use cases. Different parts address aspects like languages for the configuration of IEDs, testing of devices, data modeling, general system aspects, and the description of the ACSI, which is implemented using different communication mappings.

Originally, the IEC 61850 standard family only aimed at the field of substation automation. However, the mentioned aspects of IEC 61850 apply to automation in general and the scope of the standard family was expanded into several other fields of application. This led to several derivates of IEC 61850, like Figure 7.6 shows later on. According to this, the name of IEC 61850 was changed from "Communication networks and systems in substations" to "Communication networks and systems for power utility automation" in new parts and editions of the standard family.

Figure 7.2 illustrates the functional dependencies in the context of modeling. The physical device is virtualized and mapped onto a so called logical device. This logical device consists of so called logical nodes from the part IEC 61850-7-4. This logical device can communicate using an implementation of IEC 61850-7-2

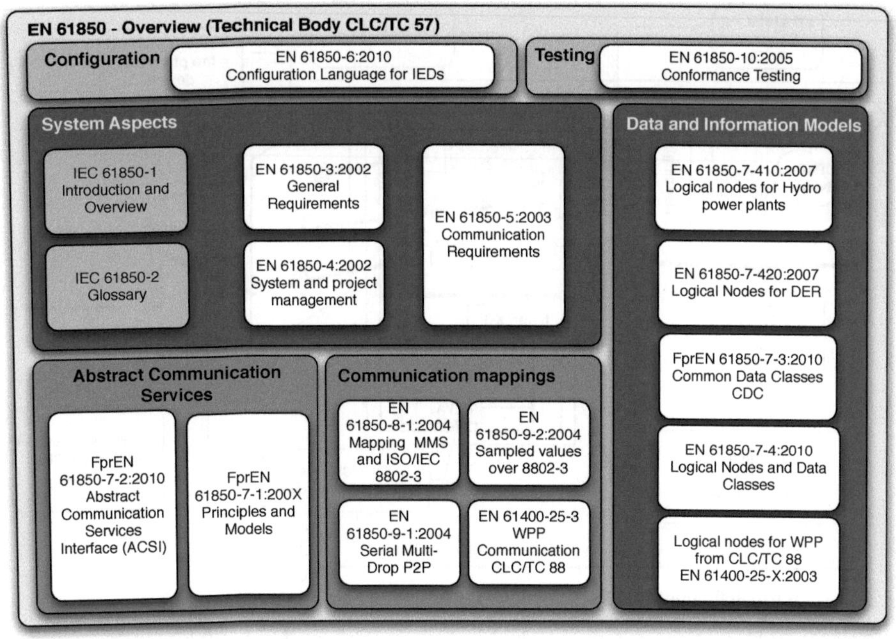

Fig. 7.1 IEC 61850 overview

conformant services, which implement the ACSI. The IEDs itself can be implemented using SCL (System Configuration Language, former Substation Configuration Language) files, which are described in IEC 61850-6. Available configuration aspects are networks, model entities, services, and single line diagrams.

The following sections will provide an overview on the three most important aspects of the standard family, the communication interface, the data modeling, and engineering aspects.

7.3 The Architecture

7.3.1 Communication Interfaces

Communication technology typically evolves over time. This is especially true for the mappings and communication base technologies. Technologies like Bluetooth, UMTS, KNX have different lifespans. While people are constantly changing their mobile phones every two years at last, the communication protocols used by that mobile phones can change, too. The life-span of substation and substation automation equipment is much longer than with consumer electronics. Starting from its erection to the end of life and decommissioning, a typical timespan is like 30

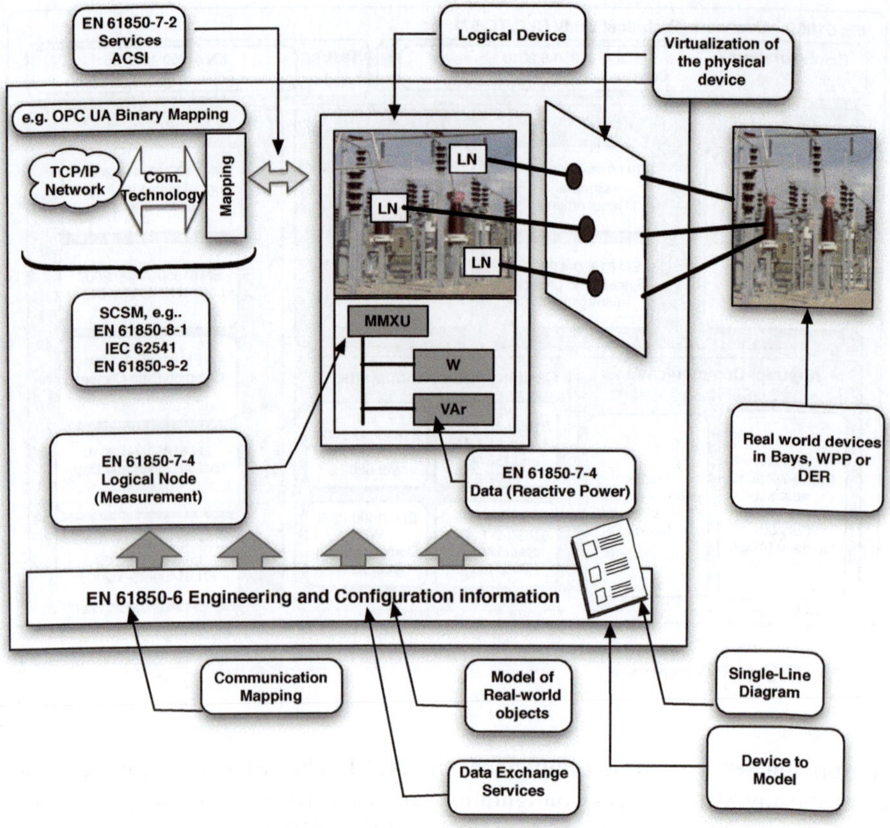

Fig. 7.2 IEC 61850 logical model

years [17]. In addition to this, features, i.e. functionality seldom changes over time. Standardization therefore has to focus on domain specific object and data models, i.e. parts of functionalities, which are common like switches, control parts, or protection equipment and not on communication technologies. The data exchanged based on those models has to cover certain meta data like time stamps, quality data attributes like validity, source, etc., which are needed by a typical SCADA (Supervisory Control and Data Acquisition) system for the safe and dependable operation of the electric power grid. Access to this data and the exchange is standardized. In order to properly address this highly relevant architectural dimension, domain-specific long-term aspects like data models and functions have been decoupled from the communication stack according to the ISO/OSI layers. This provides for easier future extension with new sub-parts while currently only the state-of-the-art based on MMS (Manufacturing Message Specification) and TCP/IP (Transmission Control

Protocol/Internet Protocol) using optical media is brought into standardization. This saves investments into the base technology because the communication stack can be exchanged with later technology without influences on the functionality. Only the communication mapping itself must be re-engineered and mapped.

7.3.2 *Modeling According to IEC 61850*

To cover basically all the modeling and communication requirements, all functional aspects of the standard family have to be drilled down to smaller attributes and objects, the logical nodes (Logical Node, LN), which communicate with each other in order to properly exchange all the needed data for day-to-day operations. The part-of relation of LNs towards certain physical devices and control levels is not standardized, which means this is really vendor or OEM (Original Equipment Manufacturer) dependent, unless certain profiles exist like, e.g., in the scope of IEC 61850-7-420 for DER (Distributed Energy Resources). Within this part, the needed LNs and attributes are "pre-choosen" for CHP (Combined Heat and Power), DER, fuel cells etc. This provides the possibility to use the LNs according to the system engineering philosophy of the utility. Multiple instances of a node can be used within a device. The functional model within the device is implemented using software. It is supported by a device model (the so called physical device, PHD) which covers all the common information about a device like name, location, vendor, manufacturer, etc. LNs are aggregated to so called logical devices (LD), the general information about the device is summarized in the logical node LPHD (Logical Node Physical Device). A schematic mapping of the general structure of a logical node is depicted in Figure 7.3 where further explanations towards IEC 61850 can be derived from.

7.3.3 *Engineering According to IEC 61850*

The data model including all the used optional attributes, their relations to the logical nodes of the physical device, the dedicated communication link, the functional mapping to the field equipment etc. can be documented for a substation in a so called single-line diagram. This configuration data is provided to the device in form of an XML-file (Extensible Markup Language) in the SCL format. Different files with different extensions for several purposes can exist. Ideally, those files can be exchanged between the engineering tools from various vendors and should, in addition, improve the engineering process. This should make for easier maintenance and extensibility of the substation systems on a larger timescale. All the needed communication interfaces of the incorporated IEDs are provided by the devices to itself, the station engineer, or, in case this is not possible, by the device manufacturer his-self on a disc or database. According to the corresponding single-line diagram, the needed functionality is added and afterwards the corresponding communication links are established. The appropriate generated file can be loaded onto the device (most of the time, unfortunately, through a proprietary interface) to engineer all the

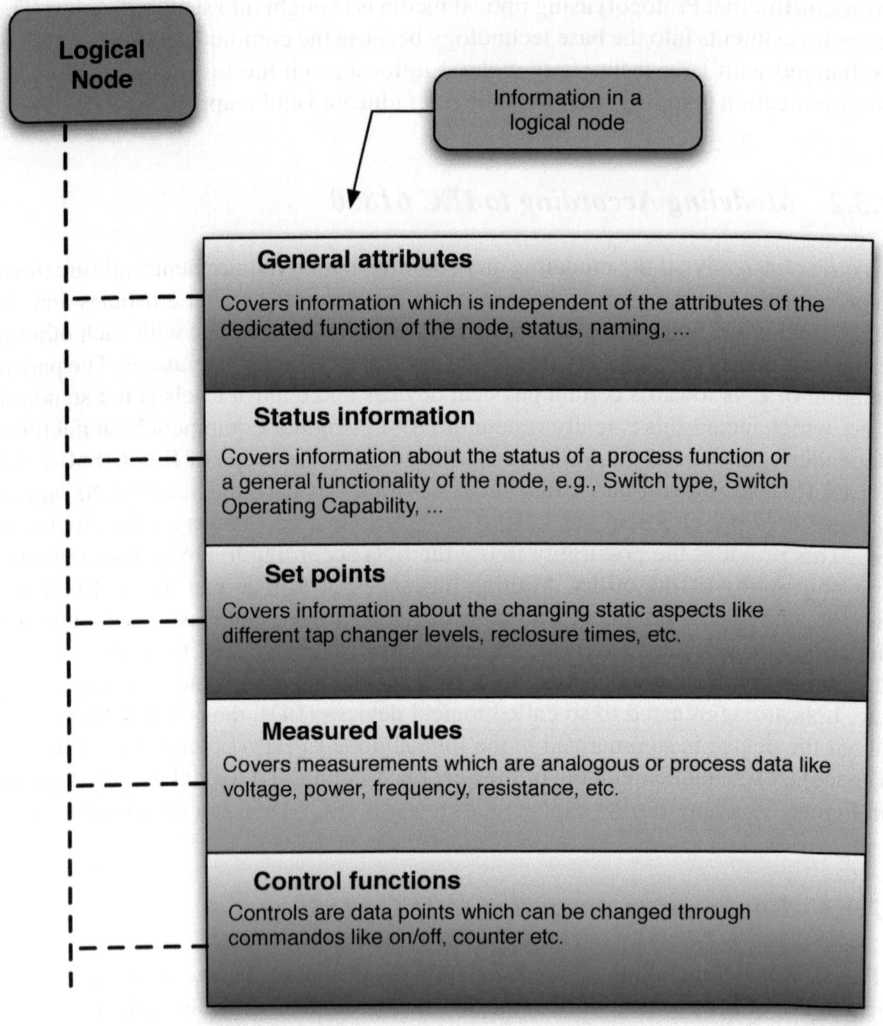

Fig. 7.3 IEC 61850 logical model: logical node example

required functionality. All IEDs should have (if they are compliant to IEC 61850-6) the ability to import and export SCL files for the engineering process.

7.4 Parts of the Standard Family

Within this section of the chapter, the sub-parts of the standard family as already depicted in Figure 7.1 will be briefly introduced.

7.4.1 61850-1 – Communication Networks and Systems in Substations – Part 1: Introduction and Overview

The sub-part IEC 61850-1: Introduction and overview [3] of the family covers a short overview on the whole standard family and a basic introduction. It covers, in addition, the detailed history of the development of the standard family, the overarching goal when it was created, an overview on the basic concepts regarding (data) modeling, communication interfaces, engineering processes, and the overall document structure of the IEC 61850.

7.4.2 61850-2 – Communication Networks and Systems in Substations – Part 2: Glossary

The part IEC 61850-2: Glossary [4] focuses on the fact that within the standard family, wording and knowledge from different technical disciplines has to be combined and harmonized, e.g., in terms of wording. Different terms from substation automation, information technology, and communications are used in context. In order to facilitate a better understanding, this part -2 harmonizes along the IEC 60050 the most important terms needed to understand the standard family.

7.4.3 61850-3 – Communication Networks and Systems in Substations – Part 3: General Requirements

The part IEC 61850-3: General requirements [1] defines general requirements in the scope of substation automation and substation systems mainly under the aspect of having the systems work in the field under no controlled environmental conditions, which has a strong impact on the requirements towards communication in terms of dependability. The part therefore provides an overview on further needed standards and pre-conditions to make IEC 61850 work.

7.4.4 61850-4 – Communication Networks and Systems in Substations – Part 4: System and Project Management

The part IEC 61850-4: System and project management [2] focuses on the very important aspect of ensuring a meaningful and canonical engineering process in order to enlarge the trust between the vendors of substations on the one hand and on the other hand the users of those systems. In terms of the communication scope, all the needed project management issues and instructions for applying IEC 61850 in a project are covered. In addition, requirements are documented in form of recommendations.

7.4.5 61850-5 – Communication Networks and Systems in Substations – Part 5: Communication Requirements for Functions and Device Models

In the part IEC 61850-5: Communication requirements for functions and device models [5], the idea of having a separation of concerns between the communication models for the functions of the field devices and the current state-of-the-art in terms of the corresponding communication mappings is discussed. The main focus of the standard family is to facilitate the use of a certain function, not the communication itself is in the very focus. By defining the functionality and the domain use cases for this functionality, the use cases identified can also cover the related requirements to those use cases. Those requirements are structured and documented in this part.

7.4.6 61850-6 – Communication Networks and Systems for Power Utility Automation – Part 6: Configuration Description Language for Communication in Electrical Substations Related to IEDs

Part IEC 61850-6: Configuration description language for communication in electrical substations related to IEDs [9] covers the definition of the SCL for IEC 61850 compliant systems. The SCL is based on the very assumption that interoperability is mainly influenced by the need of a system integrator to make systems from different vendors work seamlessly together using a number of software tools for engineering. In order to properly support this system integration, a meaningful process is required to document all device functions and communication interfaces in a formal manner. Figure 7.4 provides an overview on the integration of the SCL meta-model into the CIM domain ontology (in version 12, deprecated) whereas this is mainly a loose coupling between classes using associations and, in fact, no real integrating harmonization.

7.4.7 61850-7-1 – Communication Networks and Systems for Power Utility Automation – Part 7-1: Basic Communication Structure – Principles and Models

Based on the requirements derived from *IEC 61850-5: Communication requirements for functions and device models*, part IEC 61850-7-1: Basic communication structure - Principles and models [13] defines an object-oriented data model as well as a service-model for communication with IEC 61850-compliant field devices. This part of the standard family introduces, in addition, the overarching modeling paradigm.

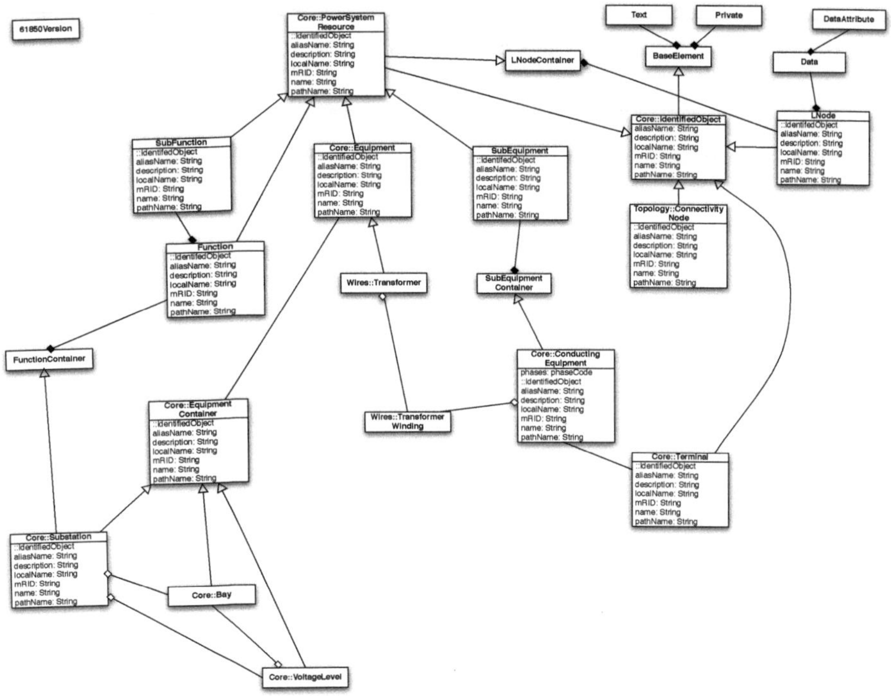

Fig. 7.4 IEC 61850 UML (Unified Modeling Language) SCL meta-model diagram from CIM 12rev0

7.4.8 *61850-7-2 – Communication Networks and Systems for Power Utility Automation – Part 7-2: Basic Information and Communication Structure – Abstract Communication Service Interface (ACSI)*

The part IEC 61850-7-2: Basic information and communication structure – Abstract communication service interface (ACSI) [10] focuses on the aspect that, in order to achieve proper interoperability according to the IEEE definition, in addition to the semantic interoperability based on a common and standardized data definition, also the interface aspect to gain access to this data must be standardized in a syntactical way using standardized data services. Therefore, this sub-part mainly covers the definitions for the common and domain-specific services. Figure 7.5 provides an overview on the ACSI and its corresponding mappings onto current communication technologies. The ACSI defines, on a conceptual layer, a platform-independent interface which can be implemented using different communication protocols. Those protocols are platform-dependent as well as application-specific and may become outdated. Currently, five different communication mappings are defined for IEC 61850 and its derivates.

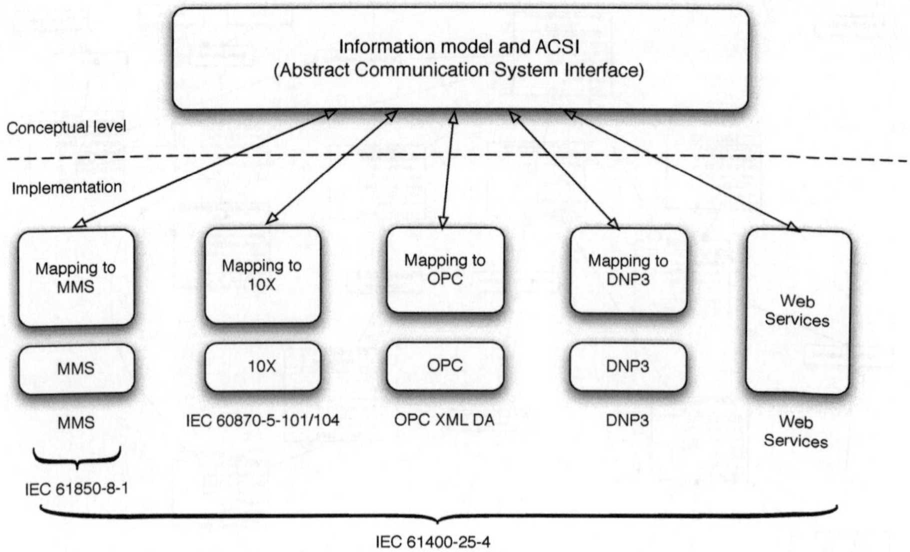

Fig. 7.5 IEC 61850 ACSI and mapping onto standards

7.4.9 61850-7-3 – Communication Networks and Systems for Power Utility Automation – Part 7-3: Basic Communication Structure – Common Data Classes

The part IEC 61850-7-3: Basic communication structure – Common data classes [15] combines all Common Data Attributes (CDA) to Common Data Classes (CDC), which are used in the part IEC 61850-7-4 of the standard family. This aggregation makes for easier understanding for developers of software as well as for the interested reader of the standard family trying to find needed functionality.

7.4.10 61850-7-4 – Communication Networks and Systems for Power Utility Automation - Part 7-4: Basic Communication Structure – Compatible Logical Node Classes and Data Object Classes

In part IEC 61850-7-4: Basic communication structure - Compatible logical node classes and data object classes [16], the exchanged data and data models are further defined based on the CDCs and standardized using the LNs. With a standardized and canonical rule set, real-world entities are abbreviated and four letter acronyms as class names are created. The first letter represents the LN group, the remaining three letters abbreviate the class of the LN, e.g., LPHD for system LNs (L) group – physical device information (PHD), AVCO for automatic control (A) group

– voltage control (VCO), and MMTR for metering and measurement (M) group – metering (MTR). The abbreviations are standardized in a manner that domain experts can still easily derive the original semantics from a quick look and provide an easy access to the object-oriented modeling and decomposition paradigm. The domain engineer can quickly go through the model and identify the needed nodes for his application as a profile. If nodes are missing, extension rules and generic nodes are given. Just like for the parts IEC 61850-7-410 and -7-420 and the IEC 61400-25 from TC 88, profiles are needed to properly address the particular different use cases with the generic modeling paradigm. Figure 7.6 covers the current derivates and profiles.

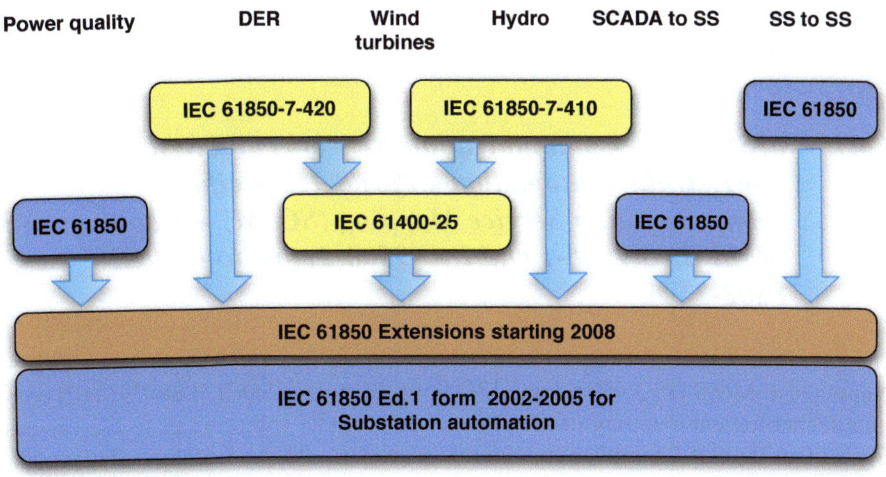

Fig. 7.6 IEC 61850 derivate models

7.4.11 61850-7-410 – Communication Networks and Systems for Power Utility Automation – Part 7-410: Hydroelectric Power Plants – Communication for Monitoring and Control

Analogous to the parts IEC 61850-7-3 and 61850-7-4, the part IEC 61850-7-410 [11] covers mainly the definitions of CDCs and LNs, which are specific to the control and communication with hydro power plants and dams. Based on the existing generic nodes from the substation automation parts, specialized models maintaining technological compatibility to the existing ones are defined and a proper profile is created. No extra restrictions are imposed and the generic semantics are still preserved.

7.4.12 61850-7-420 – Communication Networks and Systems for Power Utility Automation – Part 7-420: Basic Communication Structure – Distributed Energy Resources Logical Nodes

The part IEC 61850-7-420 [14] covers, just like the -7-410 part, analogous to the parts -7-3 and -7-4 CDCs and LNs for modeling, this time with a special focus on the control and communication with different distributed generation sources like fuel cells, photovoltaics, combustion engines, or Micro CHPs (Combined Heat and Power Plants). Again, based on the existing generic nodes from the substation automation parts, specialized models maintaining technological compatibility to the existing ones are defined and a proper profile is created. No extra restrictions are imposed and the generic semantics are still preserved.

7.4.13 61850-8-1 – Communication Networks and Systems for Power Utility Automation – Part 8-1: Specific Communication Service Mapping (SCSM) – Mappings to MMS (ISO 9506-1 and ISO 9506-2) and to ISO/IEC 8802-3

The part IEC 61850-8-1: Specific Communication Service Mapping (SCSM) – Mappings to MMS (ISO 9506-1 and ISO 9506-2) and to ISO/IEC 8802-3 [12] covers a technological mapping of the abstract data model and services to the various levels of an ISO/OSI compliant communication stack. This is usually called a communication mapping. The standard family's approach to create such a mapping is given in the parts 61850-8-1, 61850-9-1, and 61850-9-2. Part IEC 61850-8-1 defines a mapping of the so called Common Services between client (typically the human machine interface (HMI) of a SCADA) and the server (usually the IED) and the corresponding communication using GOOSE (Generic Object Oriented Substation Events) eventing and protocols between the IEDs.

7.4.14 61850-9-1 – Communication Networks and Systems in Substations - Part 9-1: Specific Communication Service Mapping (SCSM) – Sampled Values over Serial Unidirectional Multidrop Point to Point Link

The part IEC 61850-9-1: Specific Communication Service Mapping (SCSM) – Sampled values over serial unidirectional multidrop point to point link [6] specifies a mapping of analogue values over the SCSM using a uni-directional serial multidrop point to point connection, e.g., a serial communication between the electric voltage transformer and the bays in a substation, e.g., for protection purposes.

7.4.15 *61850-9-2 – Communication Networks and Systems in Substations – Part 9-2: Specific Communication Service Mapping (SCSM) – Sampled Values over ISO/IEC 8802-3*

The part IEC 61850-9-2: Specific Communication Service Mapping (SCSM) – Sampled values over ISO/IEC 8802-3 [7] defines a mapping of analogue values onto a bi-directional bus-like serial connection. Being an extension to the IEC 61850-8-1 a multi-cast of data is supported, e.g., a change of parameters of IEDs and the transmission of SCADA data and controls or even tripping is possible.

7.4.16 *61850-10 – Communication Networks and Systems in Substations – Part 10: Conformance Testing*

The part IEC 61850-10: Conformance testing [8] covers the standard conformance testing procedures, which ensure a certain degree of interoperability if certain best practices are met by the IEDs. Those non-vendor specific tests lower the risk of an error-prone integration and provide a common worldwide testing procedure, which laboratories can be certified against.

7.5 Conclusion and Outlook

The IEC 61850 standard family is an object-oriented, modern communications, engineering, and modeling solution for power utility automation, which has been established worldwide. This chapter aims to deliver the basics on modeling with IEC 61850 as well as to describe the data model, the ACSI, the reporting solution, and engineering tools including communication mappings. The target of this chapter was to mediate the fundamentals of IEC 61850 to the participant, which enables the possibility to take a future deeper look into the large standard family.

References

1. IEC: 61850 Part 3: General requirements (2002)
2. IEC: 61850 Part 4: System and project management (2002)
3. IEC: 61850-1 ed1.0: Communication networks and systems in substations - Part 1: Introduction and overview (2003)
4. IEC: 61850 Part 2: Glossary Reference (2003)
5. IEC: 61850 Part 5: Communication requirements for functions and device models (2003)
6. IEC: 61850 Part 9-1: Specific Communication Service Mapping (SCSM) Sampled values over serial unidirectional multidrop point to point link Reference (2003)
7. IEC: 61850-9-2 Ed.2: Communication networks and systems in substations - Part 9-2: Specific Communication Service Mapping (SCSM) - Sampled values over ISO/IEC 8802-3 (2005)
8. IEC: 61850 Part 10: Conformance testing Reference (2005)

9. IEC: 61850-6 Ed.2: Communication networks and systems for power utility automation - Part 6: Configuration description language for communication in electrical substations related to IEDs (2007)
10. IEC: 61850-7-2 Ed.2: Communication networks and systems for power utility automation - Part 7-2: Basic Information and Communication Structure Abstract Communication Service Interface (ACSI) (2007)
11. IEC: 61850-7-410 ed1.0: Communication networks and systems for power utility automation - Part 7-410: Hydroelectric power plants - Communication for monitoring and control (2007)
12. IEC: 61850-8-1 Ed.2: Communication networks and systems for power utility automation - Part 8-1: Specific Communication Service Mapping (SCSM) - Mappings to MMS (ISO 9506-1 and ISO 9506-2) and to ISO/IEC 8802-3 (2007)
13. IEC: 61850-7-1 Ed.2: Communication networks and systems for power utility automation - Part 7-1: Basic communication structure - Principles and models (2008)
14. IEC: 61850-7-420 ed1.0: Communication networks and systems for power utility automation - Part 7-420: Basic communication structure - Distributed energy resources logical nodes (2009)
15. IEC: 61850-7-3 ed2.0: Communication networks and systems for power utility automation - Part 7-3: Basic communication structure - Common data classes (2010)
16. IEC: 61850-7-4 ed2.0: Communication networks and systems for power utility automation - Part 7-4: Basic communication structure - Compatible logical node classes and data object classes (2010)
17. Valtari, J., Verho, P., Hakala-Ranta, A., Saarinen, J.: Increasing cost-efficiency of substation automation systems by centralised protection functions. In: 20th International Conference and Exhibition on Electricity Distribution - Part 1, CIRED 2009, Prague, Czech Republic (2009)

Chapter 8
Smart Grid Security: IEC 62351 and Other Relevant Standards

Christine Rosinger and Mathias Uslar

Abstract. Security is not only relevant for the operation of the Smart Grid as a critical infrastructure but also very important for user acceptance. This especially affects domains like Smart Metering especially in the part of privacy issues. Many different standards exist in the IEC TC57 portfolio, among them standards especially designed for end-to-end security. Additionally international security standards like ISA 99 or the NERC CIP standards were developed and will be discussed here. Furthermore this chapter describes an overview on previous attacks in the energy domain, existing solutions and security standards, and also insights on security metrics and patterns.

8.1 Introduction and Motivation

This chapter covers the IT security issue in the power system. In distinction to safety[1], security focuses the protection of security goals like confidentiality, integrity, or authenticity[2]. A subset of security is IT-security, which deals with the security of IT systems. Synonymously the term data protection is used, which additionally involves backups. Another topic in context of security is privacy, which contains the protection of the misuse of personal data [3] and appropriate realization of technical and organizational protection measures. Since this chapter handles security with focus on IT security exclusively, safety and data protection are excluded. Security concerns various layers of architectures and so security measures are necessary in every layer to ensure a holistic security. As these measures influence the

Christine Rosinger · Mathias Uslar
OFFIS – Institute for Information Technology, Escherweg 2, 26121 Oldenburg, Germany
e-mail: {christine.rosinger,mathias.uslar}@offis.de

[1] The safety of a system describes the property of accepting only functional approvable states.

[2] A definition of each of these security goals can be found in Section 8.3.1 where the glossary part of the IEC 62351 standard will be explained.

M. Uslar et al.: *Standardization in Smart Grids*, POWSYS, pp. 129–146.
DOI: 10.1007/978-3-642-34916-4_8

development of an architecture, there should be a security analysis to identify relevant security processes and measures already at the begin of the development. A retrofitting of security comes with disadvantages, like suboptimal solutions and the increasing costs and time. This approach considering security concepts in the design process is called "security by design".

It is the vision of the Smart Grid using information and communication technologies (ICT) for the electric grid to improve the efficiency, the reliability and the integration of renewable energy resources. The aspects that increase the attack and threat potential for the Smart Grid are the following:

- The deregulation leads to an increase of actors, for example electric consumers, measurement facilities, storages or distributed producers, which take part at the electricity market. Hence there is also a growth of data communication between these actors [25], which offers, if there is no security treatment before, a big gateway for attacks.
- Standard-IT is applied intensively but in addition there are a lot of domain requirements, which prevent the application of standard security measures. For example virus-checking software can block real time access of a database [5].
- There is also an increase of communication over public networks like the Internet which is another threat for the Smart Grid.

For the efficiency of the Smart Grid it is necessary to use ad hoc information, for example the current production situation. For that reason it is essential to achieve an appropriate and holistic security level for critical infrastructures like the power supply.

Another difference between Energy Control Systems and regular Office IT or for the upcoming Smart Grid is the diverse prioritization for the main protection goals confidentiality, integrity or authenticity. For Office IT the protection goal confidentiality is of highest importance, because the data that is managed at this IT is very sensitive and must not be spied. Integrity is less and availability least important. For the prioritization of the protection goals of Energy Control Systems it is the other way round, because such systems have to be available 24 hours a day and seven days a week available. This prioritization is shown in Figure 8.1.

8.2 Previous Incidents and Attack Patterns

Even without continuous ICT connections in the Smart Grid there have already been multiple attacks in this domain, which are openly published. In the last years some attacks were executed successful. For example the theft of CO_2 certificates at the emissions trading by bad identity control, caused a financial damage into the millions. Additionally, there are some different worms or viruses that attacked some industry enterprises in the energy domain, like the virus flame or the worm shamoon. Furthermore, the stuxnet worm manipulated industrial facilities for the first time and sabotaged the Iranian uranium enrichment. Especially this last attack shows that the energy domain has to be prepared for this new class of attacks.

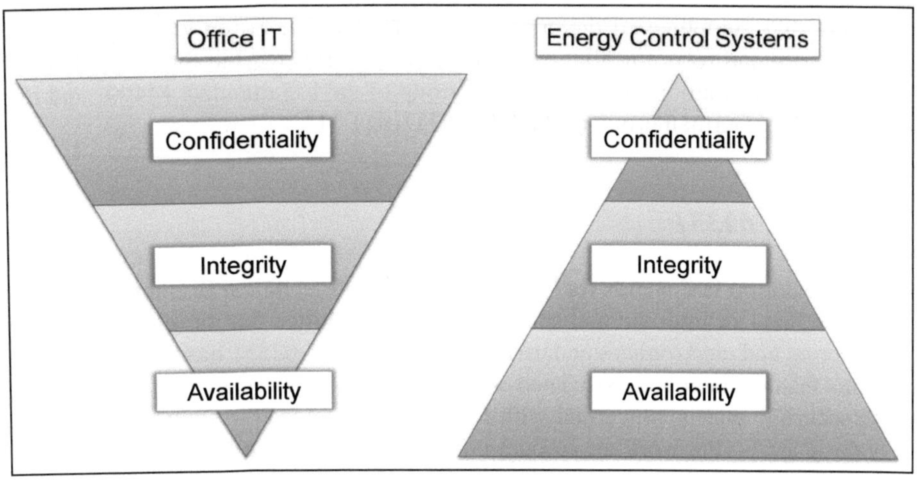

Fig. 8.1 Comparison of the prioritization of the main protection goals between Office IT and Energy Control Systems

At the moment the upcoming rollout of digital smart meter triggers currently security examinations of known smart metering systems. But also SCADA systems, which are an important element of the energy domain, or other facilities in the energy domain can get different damages. To thwart such damages and attacks and to strengthen security in the energy domain, security engineering standards are developed and applied in the Smart Grid.

8.3 Recommended Security Standards

There are already well-established security standards for different target groups and topics. The objective of all these standards, which are explained in this section, is the unification and simplification of the design process of IT security and the enhancement of the common security level. In every country different organizations are engaged in security issues. In Germany the organizations BITKOM[3] and the DIN[4] provided a German overview of common security standards without referring to the energy domain[5]. The EU project ESCoRTS (**E**uropean Network for the **S**ecurity of **Co**ntrol and **R**eal **T**ime **S**ystems) deals with the evaluation of security

[3] Engl. Federal Association for Information Technology, Telecommunications and New Media, Ger. Bundesverband Informationswirtschaft, Telekommunikation und neue Medien e.V. `http://www.bitkom.org/`

[4] Engl. German institute for engineering standards, Ger. Deutsches Institut für Normung `http://www.din.de`

[5] You can find this overview here: `http://www.bitkom.org/60376.aspx?url=Kompass_der_IT_Sicherheitstandards_final_12_11_2007.pdf&mode=0&b=Publikationen&bc=Publikationen%7cLeitf%c3%a4den`

standards and guidelines of the energy domain[6]. Another group which is familiar with security and recommendation of security standards for the Smart Grid is the Smart Grid Information Security (SGIS) group of the EU mandate M490[7] and the corresponding German group of the DKE STD1911.11[8].

8.3.1 IEC 62351

The abbreviation IEC stands for "International Electrotechnical Commission". The IEC represents an international standardization committee that develops electrical engineering and electronics standards. The standard IEC 62351 has the title "Power systems management and associated information exchange Data and communications security" and is concerned with IT security for power system management. It is designed by the working group 15 of the technical committee 57 (TC57 WG 15) and is a cross section standard included in the IEC 62357 "Seamless Integration Architecture" (SIA) [13], see also the left side of Figure 1.2 where this security standard is drawn. This standard does not cover information security management. Such security management methods can be found in the IEC 62443 or the ISO/IEC 27000 series.

The main aim of this standard is to define a secure communication infrastructure for the environment of energy management systems with end-to-end security. This also means the development of standards for security of the communication protocols defined by the IEC TC 57, specifically the IEC 60870-5 series, the IEC 60870-6 series, the IEC 61850 series, the IEC 61970 series, and the IEC 61968 series. Another objective is the consideration of end-to-end security issues. Relevant applications for the IEC 62351 in the Smart Grid domain are Energy Management Systems (EMS), Distribution Management Systems (DMS), Distribution Automation (DA), Substation (SA), Distributed Energy Resources (DER), Advanced Metering Infrastructure (AMI), Demand Response (DR), Smart Home, Storage and Electric Vehicles (EV) [24].

This standard is composed of eleven parts, with ten parts shown in figure 8.2. In this picture you can also see the correlation between the different profiles of the TC57 standards and the different parts of the IEC 62351 standard. Each part addresses different themes of the security domain. Some of the parts are not yet completed and are currently under review. The following sections give a short description of the content of the standard's content.

IEC 62351-1: Introduction and Overview The first part delivers an introduction about the challenges of IT security in the energy infrastructures and their domain specific characteristics. It also describes the considered protection goals

---

[6] See `http://www.escortsproject.eu/`

[7] See `http://www.cenelec.eu/aboutcenelec/whatwedo/technologysectors/smartgrids.html`

[8] See `http://www.dke.de/de/std/KompetenzzentrumE-Energy/Seiten/Gremien.aspx`

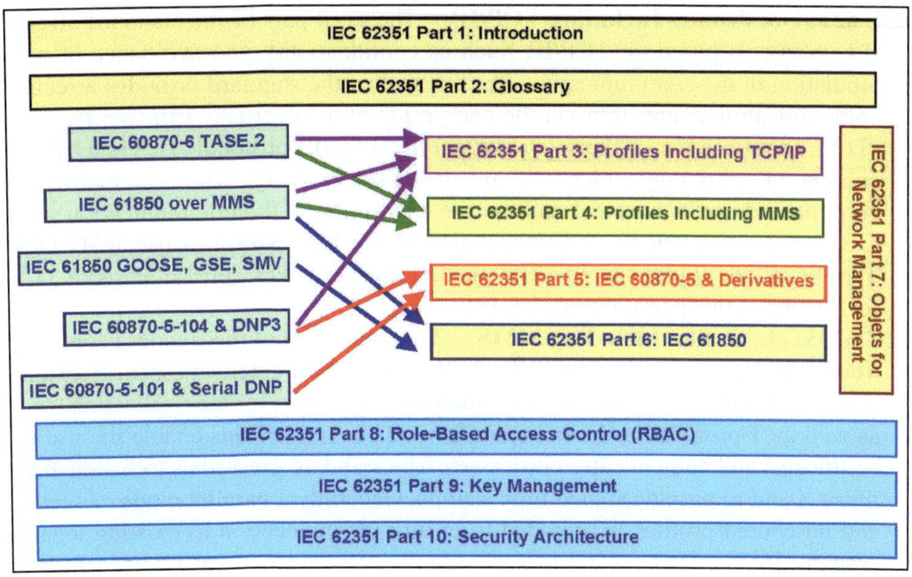

Fig. 8.2 Illustration of the correlation between the IEC 62351 and different profiles of the TC57 standards, see [8]

and the correspondent security measures. Furthermore this part shows the boundaries of this whole standard. [8]

IEC 62351-2: Glossary of Terms The second part defines terms that are used in this whole standard like a glossary. [10]

For example one definition of security in this glossary is as follows: "Security is a condition that results from the establishment and maintenance of protective measures that ensure a state of inviolability from hostile acts or influences." There are also descriptions for essential security terms like attack, threat, vulnerability or risk. It also gives an overview of the following main protection goals which are used for security considerations or risk analyses of, e.g. systems or components:

- **Confidentiality:** The protection goal confidentiality means the prevention of unauthorized access to and theft of information. A threat of confidentiality is, e.g., eavesdropping.
- **Integrity:** The protection goal integrity means the prevention of the unauthorized modification information. A threat of integrity is, e.g., manipulation.
- **Availability:** The protection goal availability e.g. means the prevention of denial of service attacks, representing a threat of availability.
- **Non-repudiation:** The protection goal non-repudiation means the prevention of the denial of an action that took place or the claim of an action that did not take place.
- **Authenticity:** The protection goal authenticity means the prevention of masquerading.

IEC 62351-3: Profiles Including TCP/IP The third part of the standard affects IEC protocols based on TCP/IP. Such as confidentiality and avoidance of manipulation in the communication. To realize this, the standard provides a secure communication connection via the encryption protocol *transport layer security* (TLS), the successor of the *secure socket layer* (SSL) protocol. TLS is a hybrid encryption technology which implements the key exchange with asymmetric encryption and certificates and realizes the encryption of the communication with exchanged symmetric keys. This technique is well-known for its use at HTTPS. The IEC 62351-3 part specifies the use of TLS and makes optional parts of TLS for the use of this part compulsory. [9]

IEC 62351-4: Profiles Including MMS The fourth part of the standard concerns MMS based protocols (see ISO/IEC 9506). The standard delivers security extensions for this protocols, which are divided into A-profiles for the application level as well as T-profiles for the transport level. These extensions enable the use of certificates and authentication processes. The standard prescribes to handle certificates and to provide signed time stamps. Generally a parallel mode of secure and unsecured profiles is provided to ensure the connection to existing legacy systems. [6]

IEC 62351-5: Security for IEC 60870-5 and Derivatives The fifth part of the standard (IEC 62351-5) deals with security aspects of IEC 60870-5 based protocols. A authentication technique on application level is described, which uses a keyed-hash message authentication code (HMAC). The IEC 62351-3 should also be taken into account, which recommends the use of authentication on particular and critical operations. In general the technique is based on a challenge-response procedure, which has to be mapped on specific protocols. The described authentication technique uses three symmetrical keys for the authentication: One update key, which has to be pre-installed on all devices, and two session keys. These session keys are used for ingoing and outgoing connections (monitoring and controlling) with different keys. The update key is only used for exchanging the session keys. [11]

IEC 62351-6: Security for IEC 61850 Profiles IEC 61850 based protocols are handled in the sixth part, the IEC 62351-6 standard. For the most part a HMAC (Keyed-Hash Message Authentication Code) is used for the authentication. Operation can only be conducted with correct Message Authentication Code. To repel Denial-of-Services (DoS) attacks, messages older than 2 minutes should be ignored. Altogether the standard advises against the use of encryption because of real time based requirements. [7]

IEC 62351-7: Network and System Management (NSM) Data Object Models In contrary of the aforementioned sub parts the seventh part, the IEC 62351-7, does not include a security extension, but covers the network and system management for energy systems. Therefore the standard specifies abstract data models, called Network and System Management (NSM) data objects, for the controlling and monitoring of the network itself and connected devices. The gained information from this can be used as an additional information source for intrusion detection systems. Through the surveillance of the network and the connected devices, it should be possible to recognize attacks and facilitate an early reaction. A guide to specific alarms is not part of this standard. [12]

IEC 62351-8: Role-based Access Control The eighth part of the IEC 62351 with the title "Role-based Access Control" is actually work in progress and therefore has not been published yet. The estimated publishing date should be in 2012. The standard itself describes role based access concepts for control systems. [14]

IEC 62351-9: Cyber Security Key Management for Power System Equipment The ninth part is under development and has not been finished yet. The main focus will be on the key management in the Smart Grid including the secure handling of cryptographic keys. [15]

IEC 62351-10: Security Architecture Guidelines The tenth part of the standard (IEC 62351-10) describes aspects of security for IT architectures in the context of TC 57-Standards. After the motivation of this standard it gives an overview of different security standards and where they are located in the TC 57 reference architecture. Furthermore there are several recommendations for a Generic Power Systems Architecture given to secure this architecture. An example is using a demilitarized zone for the connection of the operational critical parts of a distributed energy resource with other modules. [16]

IEC 62351-11: Security for XML Files The eleventh part of the standard (IEC 62351-11) defines security for XML files. The development of this standard started in May 2012, so this part is a new work item proposal. It shall standardize measures to secure XML-files while they are transmitted and the stored. [17]

8.3.2 IEC 62443 / ISA 99

The standard IEC 62334 "Industrial communication networks – Network and system security" is purposely structured corresponding with the ISA 99 standard. Both standards are referencing to specifications for information security in industrial automation. They are very similar to the ISO/IEC 2700x, except the domain specific focus. Furthermore both are providing an integration of security processes. Additionally a process model is presented.

The ISA (International Society of Automation) is a non-profit organization, with approximately 30.000 member and experts in this domain which offers support for difficult technical problems. The ISA 99 standard with the title "Security for Industrial Automation and Control System Security[9]" is concerned especially with information security in industrial automation systems and the protection of control systems. Generally it is a generic set of standards, which integrate best practices to define a cyber security management system. The standard itself consists of different parts as shown in Figure 8.3.

---

[9] The abbreviation IACS stands for "Industrial Automation and Control System Security".

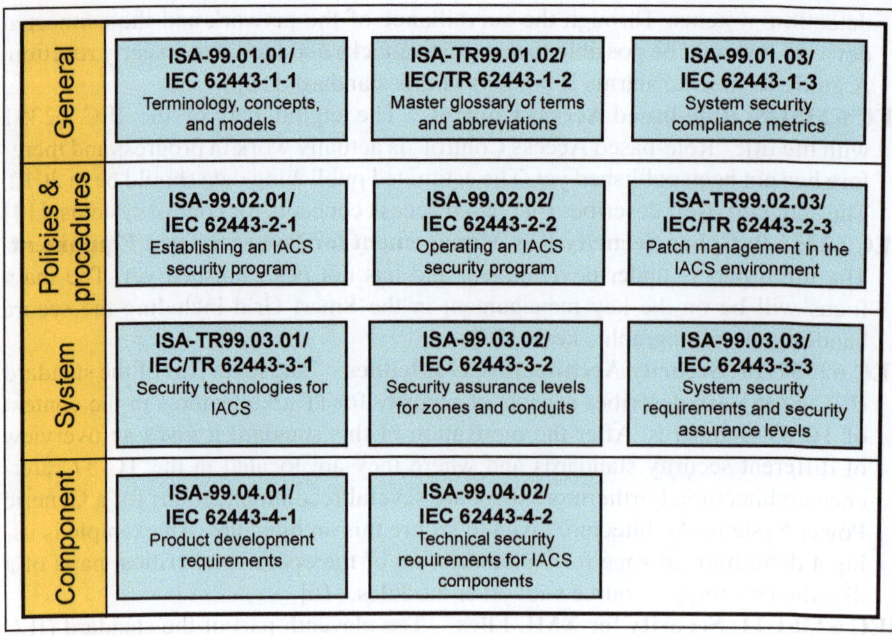

Fig. 8.3 Illustration of the structure of the corresponding standards IEC 62443 and ISA 99, based on http://isa99.isa.org/PublishingImages/ISA99_Series.png

8.3.3 NERC CIP

The "North American Electric Reliability Corporation" (NERC) is a non-profit organization aiming at improving the overall reliability of the American power infrastructure. One particular standard family they have developed is the so called NERC CIP — the "Critical infrastructure protection" program — aiming at providing means and measures to secure the supply and dependability of the power grid as a critical infrastructure. The program coordinates all efforts of the NERC: the scope of security and safety, e.g. developing standards or risk assessment. The NERC CIP Standards 002-009 have been developed in the very scope and light of the program and provide a proper framework to identify and protect assets within the critical infrastructure.

The NERC CIP parts are mandatory for the operation of power grids and power generation in the US, Canada and several parts of Mexico. In 2006, the "Federal Energy Regulatory Commission" (FERC) certified all the CIP standards and made them mandatory for securing the operation of the grid. Parts 002-1 to 009-1 have been published in June 2006 in a first version. In January 2011, all the standards have been published in the fourth edition as NERC CIP 002-4 to 009-4 [22].

All the CIP standards are based on a similar template and numbering scheme. Every part starts with an introduction, where title and numeric identifier and scope are described. In addition, the applicability is described, which authority or party has to apply the standards and where exceptions exist. The date of publishing is the last information in the introductory section. The next paragraphs focus on the individual measures and certificates which must be fulfilled in order to be compliant to the very CIP part. In addition, regional differences (e.g. Mexico or Canada) are explained.

NERC Standards CIP-002-4 through CIP-009-4 provide a cyber security framework for the identification and protection of critical cyber assets to support reliable operation of the bulk electric system. These standards recognize the different roles of each entity in the operation of the bulk electric system, the criticality and vulnerability of the assets needed to manage bulk electric system reliability, and the risks to which they are exposed. The following itemization shortly presents the main parts of the CIP standards. All these standards should be read as part of a group of the numbered standards CIP-002 through CIP-009.

- **CIP-002: Cyber Security – Critical Cyber Asset Identification**
  Standard CIP-002 requires that critical assets have to be identified to ensure reliability of the bulk system. These critical assets are assessed by risk based estimations.
- **CIP-003: Cyber Security – Security Management Controls**
  Standard CIP-003 requires that responsible entities have minimum security management controls in place to protect critical cyber assets.
- **CIP-004: Cyber Security – Personnel and Training**
  Standard CIP-004 requires that personnel having authorized cyber or authorized unescorted physical access to critical cyber assets, including contractors and service vendors, have an appropriate level of personnel risk assessment, training, and security awareness.
- **CIP-005: Cyber Security – Electronic Security Perimeter(s)**
  Standard CIP-005 requires the identification and protection of the electronic security perimeter(s) inside which all critical cyber assets reside, as well as all access points on the perimeter.
- **CIP-006: Cyber Security – Physical Security of Critical Cyber Assets**
  Standard CIP-006 is intended to ensure the implementation of a physical security program for the protection of critical cyber assets.
- **CIP-007: Cyber Security – Systems Security Management**
  Standard CIP-007 requires responsible entities to define methods, processes, and procedures for securing those systems determined to be critical cyber assets, as well as the other (non-critical) cyber assets within the electronic security perimeter(s).
- **CIP-008: Cyber Security – Incident Reporting and Response Planning**
  Standard CIP-008 ensures the identification, classification, response, and reporting of cyber security incidents related to critical cyber assets.
- **CIP-009: Cyber Security – Recovery Plans for Critical Cyber Assets**
  Standard CIP-009 ensures that recovery plans are put in place for critical cyber assets and that these plans follow established business continuity and disaster recovery techniques and practices.

8.3.4 BDEW-Whitepaper

"Requirements for Secure Control and Telecommunication Systems" is a document published as a Whitepaper[10] from the German "Bundesverband der Energie- und Wasserwirtschaft e.V." (BDEW) (Engl. German Association of Energy and Water Industries) in 2008, see [2]. The objective was to support companies in the energy domain by providing fundamental security requirements for new acquisition and new product development of corresponding systems, but also for the revision of existing systems. The requirements offer a appropriate protection to control and telecommunication systems in the energy domain in every day use. In addition the consequences of corresponding threats should be minimized, and the further business operations should be maintained or respectively should be restored as fast as possible. Because of the scope and shortness of the Whitepaper, it does not include implementation recommendations, but references to the respective paragraphs in the ISO/IEC 27002 standard, which includes these recommendations.

After the preamble where, e.g., the goal and the scope are clarified, the section "Requirements" enumerates different requirements. First general requirements of a secure system design, like the Minimal-privileges/Need-to-know principle or the Defense-in-depth principle are promoted. Further, requirements regarding patch management, encryption techniques and exact documentation are given. The handling of permanent system hardening, anti virus software or user authentication is defined in the second section "Base System". The third section, called "Networks/-Communication", gives recommendations regarding protocols and technologies to apply for a secure implementation. For example vertical and horizontal network segmentation is recommended by separating the zones with firewalls, filtering routers or gateways. Also in this section requirements for maintenance processes are illustrated. The fourth section with the title "Application" describes concepts of a secure user account management with e.g. role-based access models. Furthermore there are different references for example for application protocols or web-applications. The fifth section illustrates the development, testing, and rollout routines. Skilled staff should be appointed which utilize known standards, guidelines, and test scenarios. Furthermore secure update, maintenance processes, configuration, and change management have to be provided. The last section describes requirements for backup, recovery, and disaster recovery. In addition all these requirements should be documented for the traceability and the recommendations should be implemented only if they are appropriate.

8.3.5 ISO/IEC 27000 Series

The ISO/IEC 27000 is a generic security standard, which describes a general approach for information security management systems (ISMS). Actually the ISO/IEC

---

10 Whitepapers give an overview of different specific issues, like IT topics. They are furthermore a short paper and hence intuitive and comprehensible.

27000 series is a set of international standards for the general information security aspects. Additionally there are domain specific standards which are describing approaches of ISMS especially for the domain, like the IEC 62443.

Figure 8.4 gives an overview of the ISO/IEC 27000 series. Currently in this series there is a domain specific specification under development which is called DIN SPEC 27009 or ISO/IEC DTR 27019 and has the title "Information security management guidelines based on ISO/IEC 27002 for process control systems specific to the energy industry". This standard gives its users domain specific recommendations for information security management processes. For example it gives advices for asset management, human resource security, physical and environmental security or communications and operations management. It will be probably published in 2013.

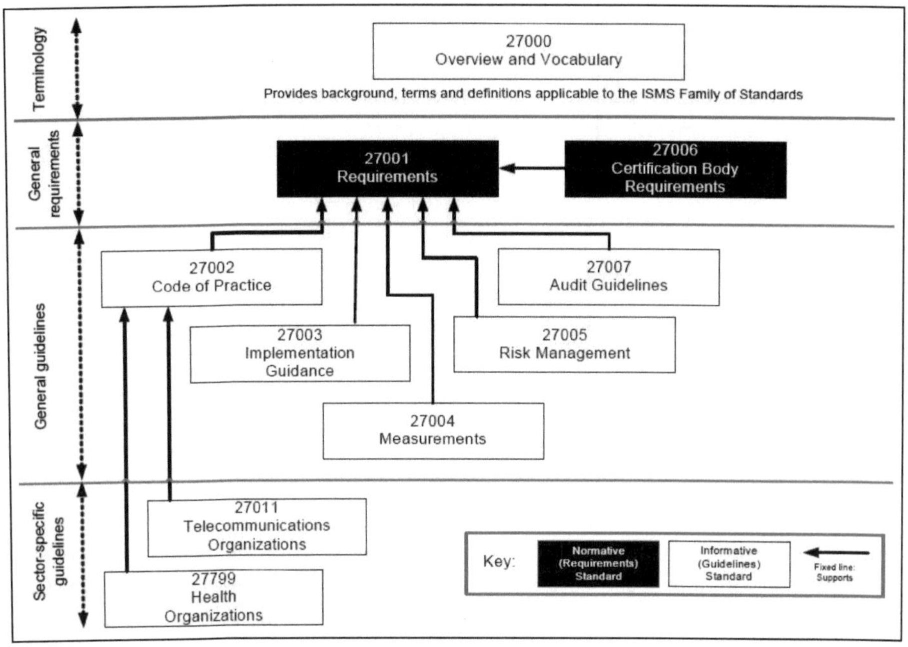

Fig. 8.4 Overview of the ISO/IEC 27000 series, see [20]

8.3.6 Protection Profile for Smart Metering

The multipart standard ISO/IEC 15408 defines criteria, which are referred as the Common Criteria (CC), to be used as the basis for evaluation of security properties of IT products and systems[11]. The CC creates comparability between the results of independent security evaluations. It does so by providing a common set of

[11] See `http://www.commoncriteriaportal.org/`

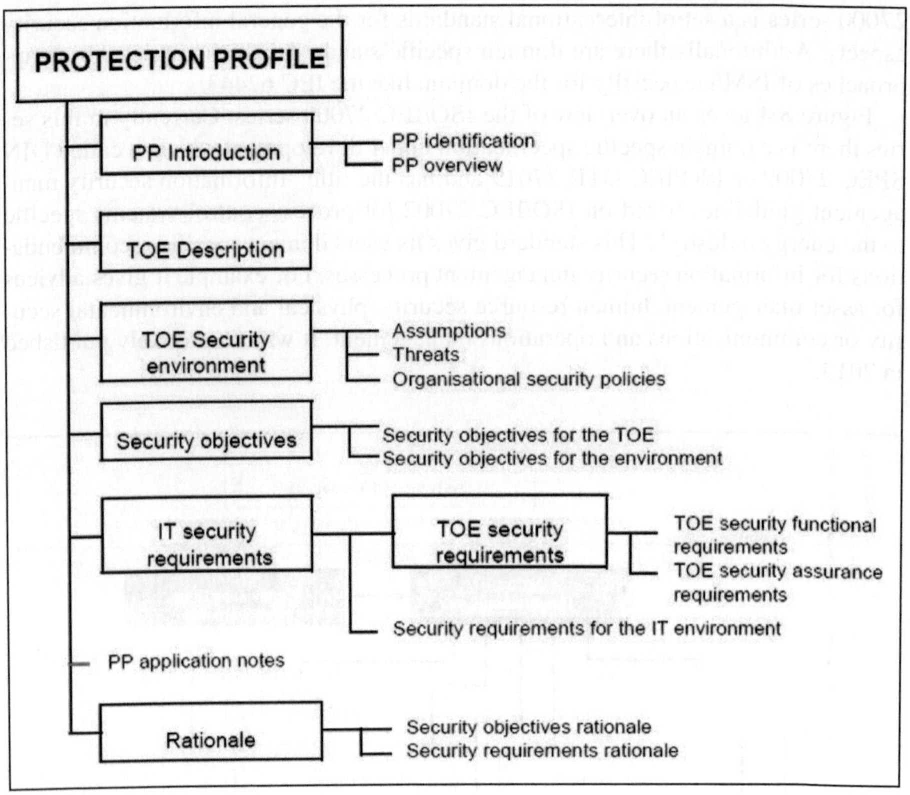

Fig. 8.5 The structure of a protection profile, see [19]

requirements for the security functions of IT products and systems and for assurance measures during a security evaluation. The evaluation process establishes a level of confidence expressing that the security functions of such products and systems and the applied assurance measures, applied to them, meet these requirements. The evaluation results may help consumers to determine whether the IT product or system is secure enough for their intended application and whether the security risks implicit in its use are tolerable. Figure 8.5 shows the structure of a protection profile. In 2011 the development of a protection profile (PP) for smart metering with the title "Common Criteria Protection Profile for the Gateway of a Smart Metering System" was begun by the German Federal Office for Information Security (in German: Bundesamt für Sicherheit in der Informationstechnik, BSI). This PP defines security requirements for the communication of metering systems, which shall also be used for evaluation of installed systems.

As shown in Figure 8.6 the BSI divided the target of evaluation (TOE) in two different parts: the gateway and the security module. The gateway "serves as the communication component between the components in the local area network (LAN) of the consumer and the outside world" [4] and the security module is used as "a

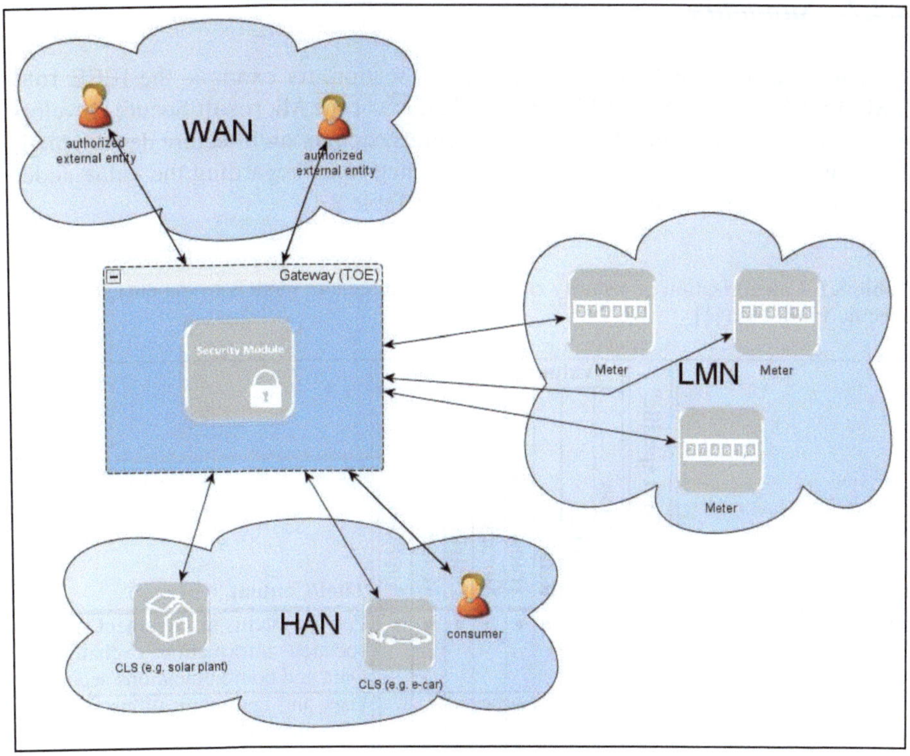

Fig. 8.6 Illustration of the components in the PP of the gateway of a smart metering system, see [4]

cryptographic service provider and as a secure storage for confidential assets" [4]. Based on this split-up, the Protection Profile is also divided into different parts: the Protection Profile for the Gateway of a Smart Metering System, the Protection Profile for the Security Module of a Smart Metering System and the technical guideline TR-03109 with additional requirements mainly concerning cryptographic methods[12].

Also shown in Figure 8.6 there are three different networks: Wide Area Network (WAN), Home Area Network (HAN) and the Local Metrological Network (LMN) for the actual meters. The figure also includes important entities. The consumer and his Controllable Local System (CLS) are in the HAN, where CLS can be producers like a solar panel or controllable loads, e.g., refrigerators. The LMN is a network just for meters, but the detailed architecture is not predetermined, so it is also possible that a meter is included within the smart meter gateway in one device. The security module is typically allocated within the gateway as it can be realized by a smart card. Authorized staff like a gateway administrator are located in the WAN and can communicate with the gateway via a secure channel (Transport Layer Security (TLS)).

[12] See `https://www.bsi.bund.de/DE/Themen/SmartMeter/smartmeter_node.html`

8.3.7 Summary

In addition to the standards presented in this section, for example the IEEE 1686 [18], AMI-SEC [26], NISTIR 7628 [25], ITIL[13], or the Microsoft Secure Development Lifecycle[14] may be of interest for secure system or architecture development. The standards described in this section are categorized regarding the value added for domains of the power system, as shown in Table 8.1.

Table 8.1 Categorization of security standards and procedure models to the energy domain specific values, see [1]

Security Standards	Smart Grid specific?	Value Added Field: Generation	Trading	Retail	Transmission	Storage	Distribution	Metering	Application	Titel/Content
IEC 62351- 1-3, 7-11	yes	•	•	•	•	•	•	•	•	Power systems management and associated information exchange – Data and communications security
IEC 62351-4, 5	yes	•			•	•	•			Data and communications security – Profiles including MMS and security for IEC 60870-5 and derivatives
IEC 62443/ISA 99	yes	•	•	•	•	•	•	•	•	Procedure model for security of industrial automation and control systems
NERC CIP 002-009	yes	•	•	•	•	•	•	•	•	Critical Infrastructure Protection for the bulk power system
BDEW Whitepaper	yes	•	•		•	•	•	•	•	Requirements for Secure Control and Telecommunication Systems
ISO/IEC 27000 series	no	•	•	•	•	•	•	•	•	International common security standards
Protection profile for smart metering	yes							•		Common Criteria Protection Profile for the Gateway of a Smart Metering System

8.4 Security Metrics

Architectures and applications for the Smart Grid need special protection because they are part of a critical infrastructure. It is difficult to declare the security level of a system, because you can measure the absence of threats but you can not prove

13 See `http://www.itil-officialsite.com/`

14 See `http://www.microsoft.com/security/sdl/default.aspx`

the presence of security. Implemented security measures and their efficiency may also be quantified. Security metrics help to measure this efficiency and supports risk management and decision processes to quantify security.

Security metrics are according to [21] defined as follows:

Definition: *Security metrics are the servants of risk management, and risk management is about making decisions. Therefore, the only security metrics we are interested in are those that support decision making about risk for the purpose of managing that risk.*

If security metrics are neglected a vicious circle can occur. This circle begins with the unawareness regarding the current security status of an application or the whole enterprise. The suspicion of an attack leads to a phase of "panic" followed by temporary closing of the identified gaps [21]. Afterwards, security activities are ending, leading back to the first stage of the vicious circle, the unawareness. Security metrics assist to break this vicious circle and thereby they are interesting for the customer and not just for developers.

Security metrics realize measuring not modeling. Statistic data is an important source for them. With such a database the metric is able to compute a ratio. Another component of the security metric is a detailed description of its intention.

Good security metrics allow continuous measurement and easy data retrieval [21]. They should deliver objective values like absolute or percental numbers. Good ones provide also ratios with an unit like "Euro" or "Hours" and reference to a context. In contrast to that, bad security metrics use subjective statements or scales like "low" till "high".

Security metrics can refer to different organizational layers. At the top strategic layer the costs for security in relation to the overall costs of IT may be considered as an appropriate metric. At the layer of quality management security metrics regarding availability like "uptime" or "MTTR" (mean time to recovery) can be used. For the common protection firewalls, intrusion detection systems (IDS), and anti virus software are interesting as metric. An IDS can be used as a metric, if it counts the number of successful attacks.

In the context of the quality assurance it is recommended to use security metrics for implemented security solutions to observe the efficiency of a measure.

8.5 Security Patterns

Software patterns provide an abstract solution for a problem in a specific context. Special patterns for security, called security patterns provide abstract solutions for typical security problems in particular contexts. The term "security pattern" is defined as follows according to [23]:

Definition: *A "security pattern" describes a specific recurring security problem, which occurs in specific contexts, and delivers a proven generic solution.*

The documentation of security patterns is similar to the one for software patterns. There is a common description, the typical context and an abstract problem and solution specification of the pattern. Furthermore, implementation details, examples, known application domains and possible side effects are documented.

Security patterns can be classified into different fields, like risk management, access control, and secure web applications. In literature there are also diverse classifications.

For example access control affects the fields of identification, authentication, and authorization. In this area accordingly patterns of identification and authentication and patterns of authorization can be distinguished. Patterns of identification and authentication deal with the unique distinction of subjects or persons (identification) and with the analysis if these identified subjects are known in particular systems (authentication). Instances of these security patterns are password-based access control with login and secret password or certificate-based access control via PKI (public-key infrastructure[15]). Further patterns in this domain are biometric and hardware-based methods.

The pattern of password-based access control for example delivers common design guidelines for the strength, the selection or the lifetime of a password. Further guidelines affect the protection of passwords during its storage, transmission and distribution. Further information is given by [23]. One solution for securing stored passwords are irreversible hash values, which are called salted hash values.

Security patterns for authorization allocate access permissions (e.g. reading, writing) to subjects for specific system resources and monitor these authorizations.

To eliminate potential vulnerabilities at the design level of an IT architecture, it is recommended to review existing security patterns beforehand.

8.6 Conclusion

This chapter showed the importance of security measures in the Smart Grid domain. It was explained that *"security by design"* is a significant paradigm that should be applied. Some established security standards were pointed out. These standards cover different topics and target groups in this domain. The main goal of all these standards is the unification and simplification of the design process for security and the enhancement of the common security level. The listing of security standards in Section 8.3.7 makes no claim of completeness, but illustrates a recommendation for the application of security standards in the Smart Grid. One specification called OPC UA (see Chapter 12), which was not explained in this chapter, also provides a security enhancement which is described in Section 12.6. Another paragraph describing security for the subdomain electro mobility is illustrated in Section 10.4.

Furthermore security metrics and security patterns were discussed. Metrics enable quality assurance of security solutions while patterns offer a common solution for a particular security problem.

---

[15] Further explanations for PKIs can be found in [3].

References

1. Appelrath, H.J., Beenken, P., Bischofs, L., Uslar, M. (eds.): IT-Architekturentwicklung im Smart Grid: Perspektiven für eine sichere markt- und standardbasierte Integration erneuerbarer Energien, 1st edn. Springer Gabler, Heidelberg (2012)
2. Bundesverband der Energie- und Wasserwirtschaft (BDEW): Requirements for Secure Control and Telecommunication Systems (2008)
3. Eckert, C.: IT-Sicherheit: Konzepte - Verfahren - Protokolle. Oldenbourg Wissenschaftsverlag (2011)
4. Federal Office for Information Security - Bundesamt für Sicherheit in der Informationstechnik: Protection Profile for the Gateway of a Smart Metering System - Schutzprofil für die Kommunikationseinheit eines intelligenten Messsystems für Stoff- und Energiemengen (2011)
5. Gresser, C., Kubik, S.: IT-Sicherheit für Leittechnik. Kes Nr. 1, 76 (2006)
6. IEC: 62351-4 TS Ed.1: Data and Communication Security - Part 4: Profiles Including MMS (2005)
7. IEC: 62351-6 TS Ed.1:Data and Communication Security - Part 6: Security for IEC 61850 Profiles (2005)
8. IEC: 62351-1 TS Ed.1: Data and communication security - Part 1: Introduction and overview (2006)
9. IEC: 62351-3 TS Ed.1: Data and communication security - Part 3: Profiles including TCP/IP (2006)
10. IEC: 62351-2 Ed.1: Data and Communication Security - Part 2: Glossary of terms (2007)
11. IEC: 62351-5 TS Ed.1: Data and Communication Security - Part 5: Security for IEC 60870-5 and Derivatives (2007)
12. IEC: 62351-7 TS Ed.1: Power systems management and associated information exchange - Data and communication security - Part 7: Network and system management (NSM) data object models (2009)
13. IEC: 62357 Second Edition: TC 57 Architecture - Part 1: Reference Architecture for TC 57 - Draft (2009)
14. IEC: 62351-8 Ed. 1.0 Power systems management and associated information exchange - Data and communications security - Part 8: Role-based access control (Draft) (2011)
15. IEC: 62351-9 Power systems management and associated information exchange - Data and communications security - Part 9: Cyber security key management for power system equipment, NWIP (2011)
16. IEC: IEC 62351-10 TR Ed.1: Power systems management and associated information exchange - Data and communications security - Part 10: Security architecture guidelines (2012)
17. IEC: IEC 62351-11 - Power systems management and associated information exchange - Data and communications security - Part 11: Security for XML Files (2012)
18. IEEE: IEEE 1686-2007 - IEEE Standard for Substation Intelligent Electronic Devices (IEDs) Cyber Security Capabilities (2007)
19. ISO/IEC: ISO/IEC 15408 - 1 - Information technology - Securitytechniques - Evaluation criteria for ITsecurity - Part 1: Introduction and general model (1999)
20. ISO/IEC: ISO/IEC 27000 Information technology - Security techniques - Information security management systems - Overview and vocabulary (2009)
21. Jaquith, A.: Security Metrics: Replacing Fear, Uncertainty, and Doubt. Addison-Wesley Professional (2007)
22. NERC: NERC CIP-002-4 bis CIP-009-4 Cyber Security (2004)
23. Schumacher, M., Fernandez-Buglioni, E., Hybertson, D., Buschmann, F., Sommerlad, P.: Security patterns: integrating security and systems engineering. John Wiley & Sons (2006)

24. SMB Smart Grid Strategic Group (SG3): IEC Smart Grid Standardization Roadmap (2010)
25. The Smart Grid Interoperability Panel Cyber Security Working Group: NISTIR 7628 - Guidelines for Smart Grid Cyber Security, vol. 1-3 (2010)
26. UtiliSec Working Group (WG) and AMISEC Task Force (TF): AMI Security Profile (2009)

Chapter 9
Testing in the Smart Grid: Compliance, Conformance and Interoperability

Robert Bleiker and Michael Specht

Abstract. Testing is indispensable when implementing complex systems, but in the case of communication standards, independent testing of different systems does not suffice. Instead, interoperability tests involving all systems, that are to communicate with each other, are needed. This chapter explains, how the efforts for this tests can be minimized, what additional benefits are achievable, and what limitations exist. Additionally, testing for the major Smart Grid standard families IEC 61850 and CIM will be exemplified.

9.1 Principles of Testing

9.1.1 Why Testing Is Necessary

The IEC 61850 and IEC 61968/70 Common Information Model (CIM) standard family, that were introduced in Chapters 6 and 7 aim at reducing the effort needed to establish communication between systems by providing interfaces for syntactic and semantic interoperability. Figure 9.1 illustrates the reduction in integration distance between two systems, that can be achieved using different levels of standardized communication.

Notice that neither IEC 61850 nor CIM are to classify as plug and automate standards. But even when using a plug and automate standard, you don't automatically achieve the intended interoperability because of several reasons:

- First of all, the mentioned standard families are complex, both consist of several thousand pages.

Robert Bleiker · Michael Specht
OFFIS – Institute for Information Technology, Escherweg 2, 26121 Oldenburg, Germany
e-mail: {robert.bleiker,michael.specht}@offis.de

M. Uslar et al.: *Standardization in Smart Grids*, POWSYS, pp. 147–161.
DOI: 10.1007/978-3-642-34916-4_9

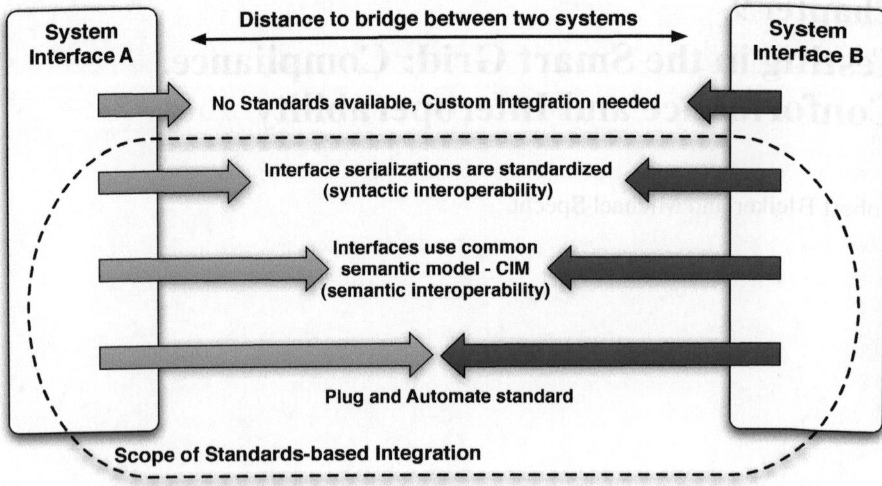

Fig. 9.1 Integration distance between two systems

- Additionally, the standard families are written by human beings, who can and sometimes will make flaws.[1]
- The standard families also are, at least partially, written in a natural language, English in case of IEC 61850, CIM and other international standards, which is ambiguous in contrast to formal languages and which is not always the writers first language.
- And after that, the standards sometimes get used or translated into other natural languages by people, whose first language also is not English.

In Software Engineering, continuous testing is an established way of handling the complexity of systems [11]. But concerning communicating systems, that may even be built by different vendors, independent testing of the systems during their development does not adress all of the problems mentioned above. Therefore, to ensure interoperability, bilateral integration tests of systems, that should communicate with each other, are needed. In a centralised scenario with only one client controlling several servers, this can be done with limited effort. If a new server is to be added to the scenario, only a test between this server and the central client is needed to ensure further interoperability. The number of tests needed already increases considerably, if the scenario does not only allow several servers, but also several clients. Adding another system to this scenario requires bilateral tests between the new device and all devices of the other type. This still gets worse, if there is no differentiation between clients and servers and every system is allowed to communicate with each other system. In this case adding a new system requires testing the communication

[1] See for example IEC 61850 technical issues at `http://www.tissues.iec61850.com`

with all other systems. This means the number of tests needed is rapidly increasing with the number of systems involved. To avoid these efforts and the associated costs, a centralized testing instance is needed. Figure 9.2 visualizes these efforts. Each line connecting two systems also represents a needed test execution.

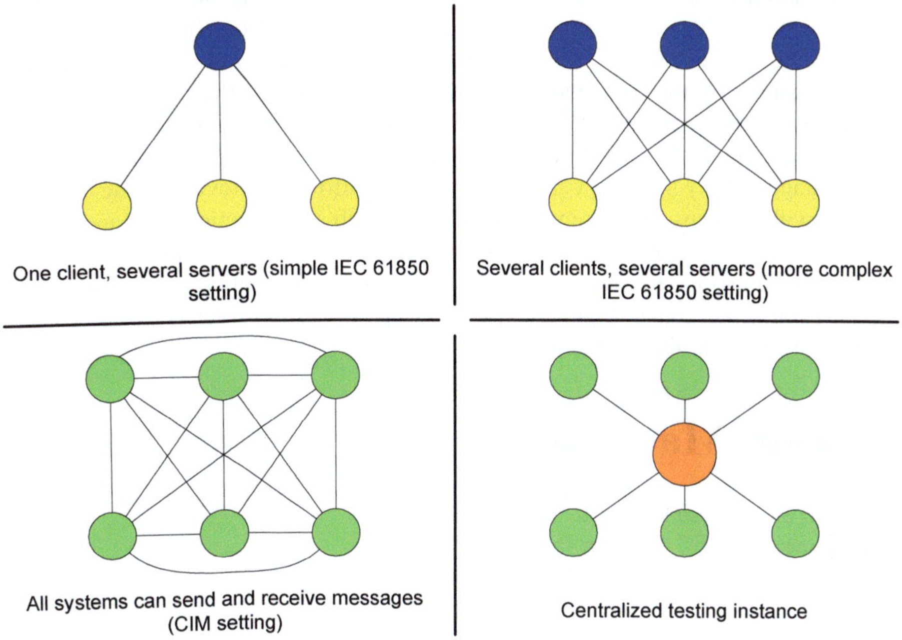

Fig. 9.2 Efforts of bilateral testing in different settings

A centralized testing instance can further be used

- for quality assurance,
- as a reference,
- to adress integration difficulties,
- to lower development efforts and
- to clarify issues of interpretation.

This centralized testing instance shall further be called testing machine.

9.1.2 Testing Requirements

Different stakeholders may have completely different requirements referring to the testing machine. The requirements described here are established aiming to be most beneficial for the general realization of a Smart Grid scenario.

The testing machine should be able to test any device, that implenents either IEC 61850 or CIM, for conformity with the correspondent standard family, or to test the data generated by that device in the same way. Thereby it is assumed, that only the device itself or data generated by that device is available, but no additional information unless the corresponding standard family specifies it to be present.

To be able to use the testing machine for continuos testing during the development of a device, the execution of a single test should not be time-consuming or expensive. To achieve this, a fully automated test is aspired, that does not need any input or actions by the user once it is started. Additionally, the tests shall be executable without the need to have a local instance of the testing machine. This is achievable because both standard families, IEC 61850 and CIM, adress the field of data exchange. They serve to enable the remote control of a device or the communication with it. Thus, the testing machine with web service interface can be used in heavily automated processes. In order to also provide comfortable remote access to the testing machine for human users, an additional web application to control the testing machine is needed. Figure 9.3 presents a possible user interface of how a test can be configured in the web application.

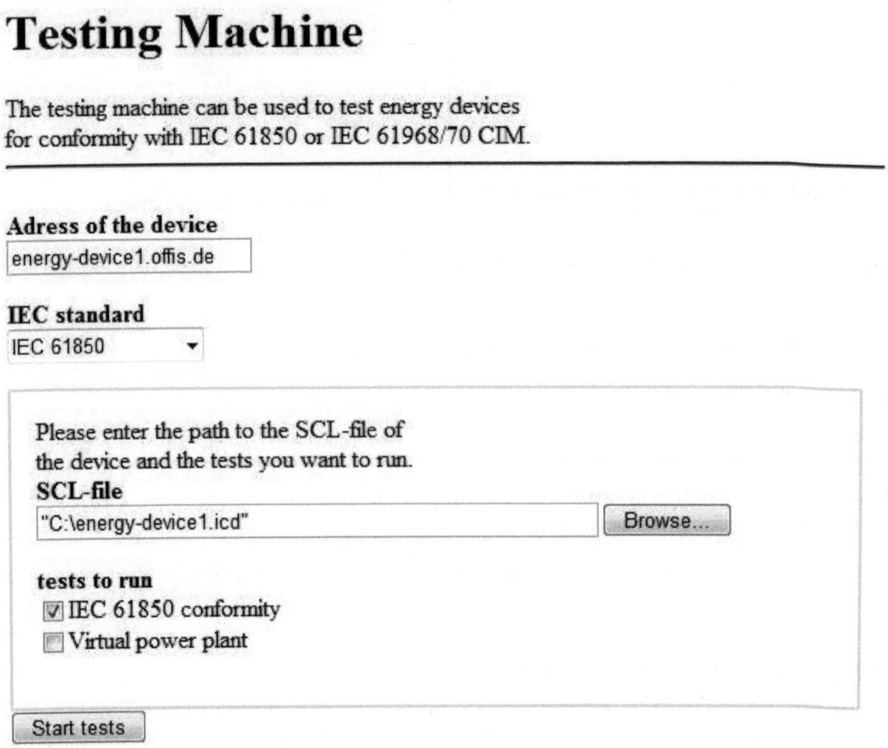

Fig. 9.3 Ordering the test of an IEC 61850 device in the web application

Because the devices to be tested may differ considerably regarding the used hardware platform and the software it runs, the use of static testing procedures is not applicable. Static testing procedures like an analysis of the source code are carried out without the execution of the operation software of the device. Using static testing procedures would imply considering all the differences in hardware and software, so in case of the desired automated testing of a wide variety of devices, it is out of scope. In addition to the effort needed for the development of a testing machine, that would be capable of handling all these differences, it can finally not be assumed, that the source code of the tested device is even available.

Furthermore the testing machine shall be easily extendible by test cases, that analyze the applicability of the tested device to be deployed in different concrete scenarios defined by the user of the testing machine himself. For example a device designed to praticipate at a local energy market may need certain data structures, that are optional in terms of standard conformity. Regarding the standard family CIM, it is also desirable to test the standardized profiles.

To execute a test case, the user has to provide all test relevant information to the testing machine. In case of a CIM message, this only includes the message itself and information about whether additional test cases, like mentioned in the paragraph above, should be executed or just the test case for standard conformity itself. To test an IEC 61850 device, the configuration file of that device is needed and an internet protocol (IP) adress or uniform resource locator (URL) to enable the testing machine to establish a communication connection to the device to be tested. While testing a CIM message does not need access to the device that generated the message at all, when testing an IEC 61850 device the communication connection is used to send different commands to the device and analyze the answers. in both cases a detailed testing log is generated and made available to the user. Figure 9.4 exemplarily shows the communication cycle in case of a human user and an IEC 61850 device. In addition it shows the different components of the testing machine needed to perform its tasks.

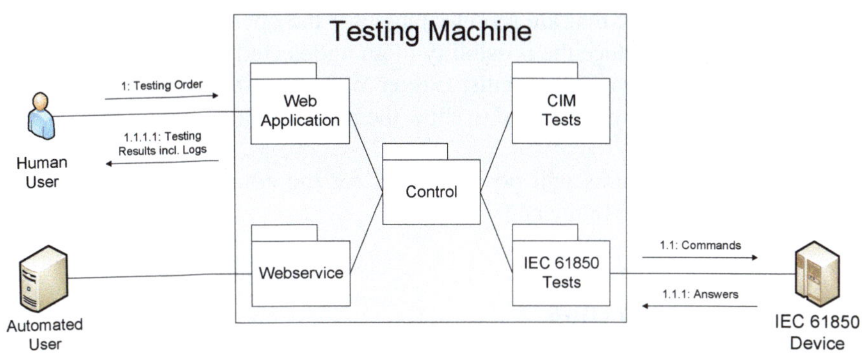

Fig. 9.4 Communication during a test and design of the testing machine

9.1.3 *Limits of Testing*

An upper limit for the possibilities of the testing machine is presented by Rices theorem [10], which states that it is impossible to automatically answer whether an algorithm, exclusively according to its input and output, has a certain property or not.

There are additional limitations for the here defined testing machine coming from the requirements that the testing machine should fulfill, since only the visible behaviour via the used communication interface and the data generated by the tested device can be observed. Other than Rices theorem premises, the testing machine does not have access to the algorithms used by the tested device. Therefore there is no possibility to observe internal procedures of the tested device and draw conclusions concerning the processing of data from that. Only input, output and the associated behavior of the tested device can be observed. Beneath the already mentioned non-applicability of static testing procedures also some dynamic testing procedures, that require information concerning internal procedures like the control flow of the tested device, can not be used. The aspired test from distance can be subject to additional limitations caused by the communication technology used. For example, the testing of real time specifications is not possible via a communication interface, that does not fulfil them.

Thus to test a device for standard conformity, only its behavior can be observed, especially the data generated by the device can be analyzed. A successfully passed test means, that the tested device acted standard conform unter the conditions of the test and hence, it is basically capable of standard conform behavior. The test can not guarantee, that the device will act standard conform under all other possible conditions. Testing the device under all possible conditions is not viable, since all combinations of possible external parameters and inner conditions would have to be taken into consideration. The effort to do this is growing exponentially with the number of the considered factors, the time of such a testing series would therefore exceed the economic life-time of the tested device. Executing the test under a selection of conditions, that are representative for the operation of the device, can however drastically reduce the possibility of an undetected fault. A well-known citation of the Dutch computer scientist Edsger Wybe Dijkstra summarizes this very well: "Program testing can be used to show the presence of bugs, but never to show their absence!" [1]

The following sections will go into detail for the conformance testing of the standard families IEC 61850 and CIM.

9.2 IEC 61850 Testing

Within the IEC 61850 standard family, part -10 deals with conformity testing for the series itself. The current first edition only covers the conformity testing of servers. In the IEC 61850 client-server-model a server is a communication device that awaits the establishment of a connection by the client. After that the client controls the server using the connection and can also terminate the connection afterwards.

A fully automated test for standard conformity, like it is aspired, can thus be realized for any IEC 61850 server, but not for any client. If a client had to be tested, the testing machine would have to take the role of the server and would for this reason not control the client, but be controlled by the client. Thereby it could not actively execute the test. On the other side, it can not be assumed that any client is prepared to automatically carry out all steps needed for a conformity test. Such an IEC 61850 client would have to be modified before testing, which is out of the scope of the desired fully automated conformity tests.

Hence this chapter focuses on the basic aspects of the automated testing of IEC 61850 servers, which make up the majority of IEC 61850 devices in most common Smart Grid scenarios. To ensure interoperability in spite of the IEC 61850 clients not being tested, it is highly recommended to build the testing machine as similar as possible to the clients to be used, ideally modify a client to act as testing machine. The following Sections 9.2.1 and 9.2.2 will explain testing of the configuration and the data model of an IEC 61850 server.

9.2.1 Configuration Testing

IEC 61850-6 specifies a description language for the configuration of devices, the System Configuration Language (SCL). It is used to describe the electrical and mechanical components of a device, the automation systems, the communication systems, and the relations between these. Additionally, it enables the possibility to exchange configuration data between different devices and configuration tools. SCL is based on version 1.0 of the Extensible Markup Language (XML).

IEC 61850-6 specifies that every standard conform device has to come with a SCL-file, which describes the current configuration of the device. In addition, it must be possible to directly configure the communication settings of the device using the SCL-file, as long as they are configurable at all. Alternatively, the device can come with a tool that generates the SCL-file from the current configuration of the device and can also modify this configuration using a SCL-file. Concerning the testing machine, the availability of a SCL-file, that completely describes the configuration of its device, can be assumed. Otherwise, the device would not be IEC 61850 conform. There are at least six different types of SCL-files defined, the IED Capability Description (ICD) is typically used as configuration file for an operational device.

To test the configuration of a device, the testing machine carries out the following test steps:

- The first test step consists of the inspection of the configuration file for syntactical correctness. In this process it is checked, whether the configuration file is a correct XML file and if the configuration file adheres to the schema for SCL-files, which is defined in part -6 of the standard family. A test, whether the data model specified in the configuration file complies with all relevant parts of IEC

61850 does not take place now. Instead, this is tested later with the data model of the device itself.

- In the next test step, the complete data model configured in the SCL file is matched with the data model of the operational device and analyzed for differences. This match includes all names, the data types, the collections of data called Data-Set and some predefined values. If parts of the data model configured in the SCL-file are missing at the device, this is classified as failure. But if the device presents a data model with additional data compared to the SCL-file, this is not classified as failure, but will only be pointed out as a note, because the data model configured in the SCL-file is completely implemented by the device.

9.2.2 Data Model Testing

After the testing of the configuration, the data model is tested. The data model has a tree-like structure, the root of such a tree is always a Logical Device. A Logical Device has several Logical Nodes, at least both Logical Nodes "LLN0" and "LPHD", which are mandatory. Every Logical Node has several Data Objects. A Data Object can have Sub Data Objects, which behave like additional Data Objects, and Data Attributes. A Data Attribute can be a simple date like a number or a character string, a more complex date like an array or a time stamp, or it can consist of several Sub Data Attributes, which all behave like a Data Attribute. IEC 61850-7-2, -7-3, -7-4, -7-410, -7-420, and IEC 61400-25-2 (Communications for monitoring and control of wind power plants – Information models) contain predefined Data Attributes, Data Objects and Logical Nodes for different purposes. Beneath the use of these predefined data structures, IEC 61850 also allows the extension of Logical Nodes, Data Objects, and Data Attributes and even the definition of completely new ones. Figure 9.5 shows part of an exemplary data model of an IEC 61850 device. Here only the sub nodes of nodes on grey background are shown.

After the previously described test steps covering the configuration of the device have been carried out, the data model should be checked. Thereby the following test steps are executed:

- The presence of the mandatory Logical Nodes "LLN0" and "LPHD" mentioned above has to be checked.
- After that, every node of the tree will be checked for whether it is a predefined data structure or not. If a node is detected, that does not fit any of the predefined data structures, it should be noted that the data model has been extended at this point.
- The predefined Logical Nodes, Data Objects and composed Data Attributes each have sub nodes, whose presence is mandatory in some cases, optional in others and also may be mandatory or forbidden depending on different conditions. The testing machine shall check, if all these specifications are followed by the data model of the device.

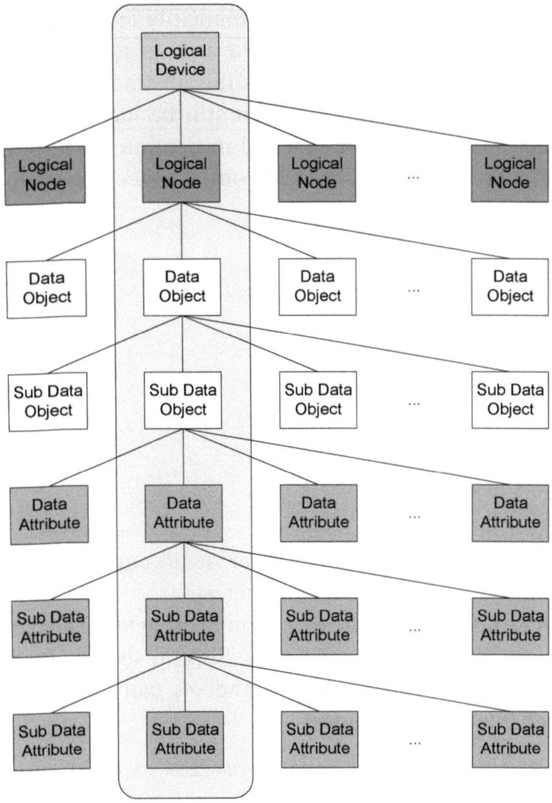

Fig. 9.5 Exemplary part of the IEC 61850 data model

- In case of Data Attributes that are not composed, it should be verified if the correct data type is used and in some cases if the range of values is kept.
- In case of a Logical Node, Data Object or Data Attribute, the order of the sub nodes is specified. This also has to be tested.

9.2.3 Further Testing of IEC 61850

To allow different technical implementations and also to be able to benefit from progress in communication technology, the IEC 61850 standard family does not directly specify the way information has to be transmitted. Instead, it defines an abstract interface and additional mappings to communication technologies with the mapping to Manufacturing Messaging Specification (MMS) being the most important one today. These mappings can be exchanged and it is also possible to extend IEC 61850 with completely new mappings. However, while testing an existing device, a specific mapping will be used and should also be tested during the whole process. Thereby the described way of automatically testing a device without any knowledge of its internal processes can make it impossible to determine whether certain types of errors are either related to the abstract interface or the specific communication mapping. However, they can still be identified as errors, only the accurate determination needs a more detailed inspection of the device.

A lot of additional testing of IEC 61850 conformance is possible in the fields of

- starting and ending communication,
- data sets,
- substitution,
- setting groups,
- reporting,
- logging,
- events,
- control,
- time,
- and file transfer.

This section will however not go into further detail concerning those fields, as it concentrates on configuration and data model testing.

The theoretical limits of testing described in Section 9.1.3 also have practical impact when testing for IEC 61850 conformity. Several specifications of IEC 61850, especially but not only from the parts -3 and -4, can not be tested in the way described in this chapter. This includes:

- specifications that refer to the design or process of building of a device or its components
- specifications regarding the mechanical structure and physical characteristics of the device
- specifications that require specific information to be present with the device, but without ordering them to be in machine readable form or even available via the communication interface
- specifications referring to how the device has to behave in case of an error
- requirements regarding real time characteristics
- specifications concerning the internal behavior of the device

Additionally to only testing conformity with IEC 61850, the testing of security aspects could also be desired. In this case, the specifications of IEC 62351-3, -4, and -6 would be of relevance. The IEC 62351 standard family is described in section 8.3.1.

9.3 Common Information Model (CIM) Testing

Conformance testing of message based communication like it is described in the Common Information Model (CIM) needs a whole different approach. The following section covers this.

9.3.1 How to Test CIM Messages

In the domain of coupling between SCADA and primary(process IT) as well as secondary IT (commercial IT), the IEC developed in the midst of 90s the so called Common Information Model CIM (IEC 61970/61968). To match different use cases, the CIM is not only developed as Energy Management (EMS)-API, but also as a domain data model for the energy supply industry, worldwide. Further introduction to the CIM can be found in Chapter 6.

The main communication between systems with CIM is done by using XML messages. One of the main problems of testing CIM compatibility is, that the CIM does not support testing like the previous mentioned IEC 61850. There is no testing mode for servers or clients, which could be used. Therefore the only way to test CIM communications without modifying client and server applications, is to test the XML message itself. This limits the testing, so that the testing of complete processes will not be available.

Finally the following tests are possible and reasonable:

1. general XML conformity
2. general CIM message structure
3. general use of CIM elements as XML tags
4. specific CIM payload with a given payload structure as schema

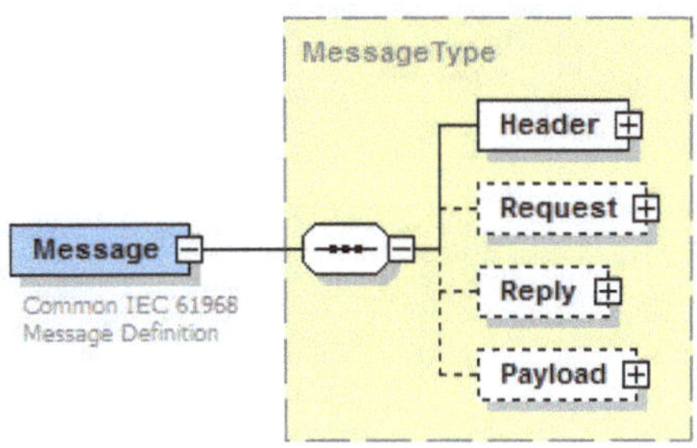

Fig. 9.6 CIM Generic Message Structure [9]

To test the first point nearly every XML API in most programming languages offers generic functions to test XML documents on general XML conformity. In most cases the XML compatibility is a requirement to load the message anyway.

The second point "general CIM message structure" uses the predefined message structure given from the IEC 61968/61970 standard families [9]. Figure 9.6 shows the minimized overall message structure, which can be roughly splitted into the header (shown in figure 9.7), additional elements like reply or request elements, and the payload. This message structure is available as an XML Schema, enabling to directly test a message regarding this schema. Similar to the first point, many XML APIs and libraries are available which offer these functionalities. With passing such

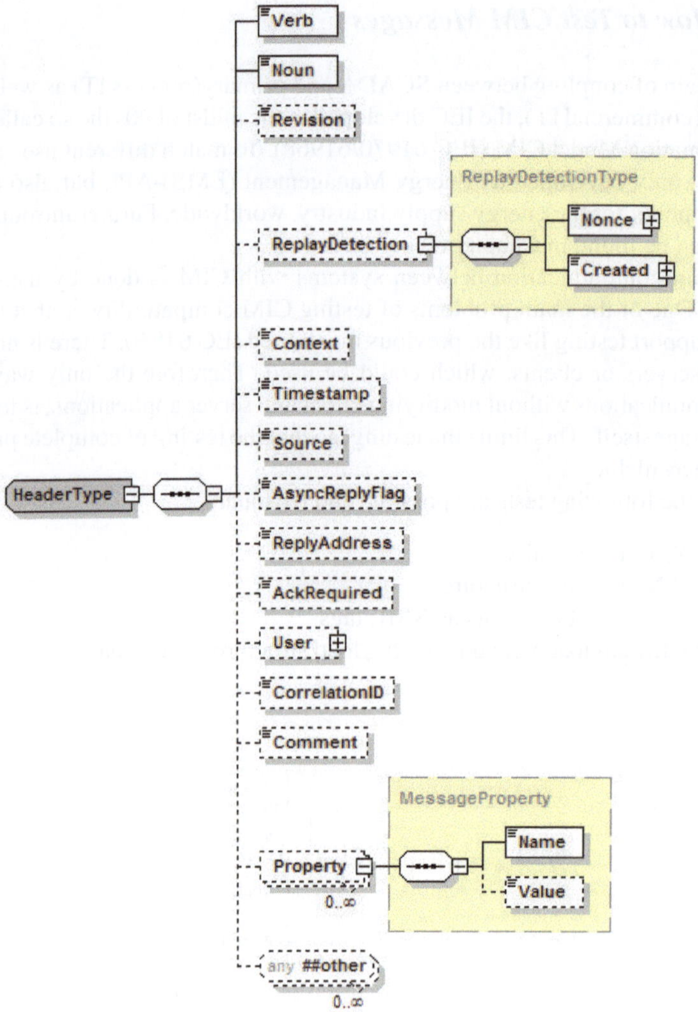

Fig. 9.7 CIM Header Message Structure [9]

a test the only parts left are the payload, which can be any element regarding the standard XML Schema.

With only the payload left, point 3 on the list should be tested. The CIM offers a digital data model in UML. This data model can be used to verify whether the tags used in the payload refer to CIM objects or not. A basic string comparison would be sufficient to answer this question.

At last the payload itself should be specified beforehand, whether by standards like the IEC 61968-9 or by experts/developers. These definitions should be digitaly available. For testing the same procedure mentioned in point 2 can be used, except the test is focused only on the payload part.

9.3.2 CIM User Group Interoperability Tests

The aim of the Interoperability (IOP) Tests is to demonstrate that different vendor systems are able to successfully exchange models using the IEC interface standard families 61970 and 61968. IOP Tests use specific CIM profiles (see Section 6.3) to perform compliance and conformity tests.

In the past, IOP testing has mainly focused on three areas:

1. exchanging power system network models for transmission (CPSM, ENTSO-E) and distribution (CDPSM) using the CIM
2. compliance and interoperability testing of the GID standards
3. exchanging messages based on the IEC 61968 standard parts (mainly metering 61968-9)

IOPTests [7, 6, 8, 4, 5, 2, 3] are organized by IEC TC 57 WG and CIMug according to industry needs and usually rely on the latest agreed combined version of the CIM. During IEC TC 57 WG meetings, dates and locations for IOP as well as test cases and CIM profiles to be tested are agreed on. Based on this through IEC WG and CIMug mailings, WG and CIMug members are invited to take part in IOP Tests. CIMug or IEC TC 57 WG membership of participants is expected but non-members might take part as well.

Mainly three types of actors can be distinguished within IOP Tests:

- **participants:** vendors who want to test their applications
- **observers:** witnessing the test
- **validation experts:** supporting the use of validation tools like for instance CIM-Spy or CIMDesk
- **organizer:** hosting and organizing the meeting as well as preparing the report

At the moment no standardized procedure model for conducting IOP-Tests is available. In the following basic steps within interoperability tests based on several descriptions of IOP-Tests are provided.[2] Individual IOP-Test may vary and change in the future as the procedures are always agreed on by the organizing WG members and the participants.

Figure 9.8 shows the general IOP test procedure.

As a prerequisite within WG meetings the latest CIM version, IOP test cases and CIM profiles which should be subject of the IOP Test have to be defined and organizational aspects (like host, dates, location, participants, and report publication) are agreed on. Then CIM XML test files have to be provided by each participant (Step 1). All provided files need to be validated against the underlying CIM profile and corrected in case of violation (Step 2). CIM validation tools like CIMSpy or CIMDesk are usually used. On basis of the validated files import and re-export

[2] See for example EPRI and ENTSO-E IOP test descriptions `http://mydocs.epri.com/docs/public/000000000001013295.pdf`, `http://mydocs.epri.com/docs/public/000000000001012494.pdf` or `https://www.entsoe.eu/fileadmin/user_upload/_library/news/CIM_IOPs_and_Roadmap_Explanatory_note.pdf`

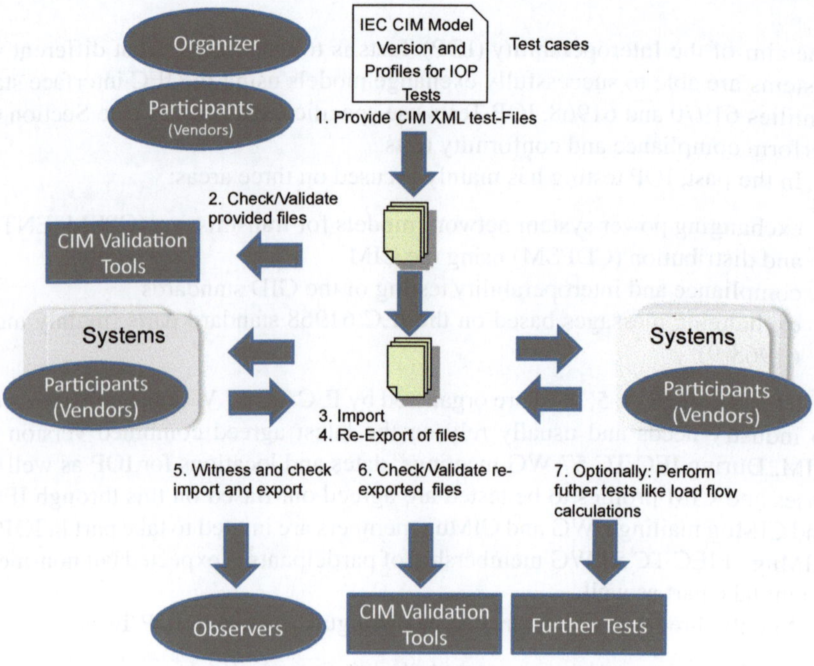

Fig. 9.8 CIMug IOP Procedures general overview [12]

(Step 3 and 4) are performed by the applications of the participants. Observers act here as witnesses to check the import and re-export. Re-exported files will usually be validated again with CIM validation tools. Participants may decide to only take part in dedicated test cases. Regarding on the type of IOP tests and data exchanged, further tests might be conducted. For example when dealing with the exchange of network models, load flow calculations might be performed and compared.

References

1. Dijkstra, E.W.: Notes on structured programming (1970)
2. EPRI: Report on the Common Information Model (CIM) Extensible Markup Language (XML) Interoperability Test #1 (2001)
3. EPRI: Report on the Common Information Model (CIM) Extensible Markup Language (XML) Interoperability Test #2 (2001)
4. EPRI: Report on the Common Information Model (CIM) Extensible Markup Language (XML) Interoperability Test #3 (2001)
5. EPRI: Report on the Common Information Model (CIM) Extensible Markup Language (XML) Interoperability Test #4, Cim (2002)

6. EPRI: Report on the Sixth Control Center Application Program Interface (CCAPI) Interoperability Test; The Power of the Common Information Model (CIM) and Generic Interface Definition (GID) to Exchange Power System Data (2004)
7. EPRI: CIM-XML Interoperability Including CIM-Based Tools Test: The Power of the Common Information Model (CIM) to Exchange Power System Data (2008)
8. EPRI: Smart Meter Information Interoperability Test (2010)
9. IEC: 61968-100 (Draft): Application integration at electric utilities - System interfaces for distribution management - Part 100: Implementation Profiles for IEC 61968 (2011)
10. Rice, H.: Classes of Recursively Enumerable Sets and Their Decision Problems. Transactions of the American Mathematical Society 74(2), 358–366 (1953)
11. Sommerville, I.: Software Engineering, 9th edn. Addison-Wesley, Amsterdam (2010)
12. Uslar, M., Specht, M., Rohjans, S., Trefke, J., González, J.M.: The Common Information Model CIM: IEC 61968/61970 and 62325 - A Practical Introduction to the CIM. Springer (2012)

Chapter 10
Standards in the Electro Mobility Domain—Vehicle 2 Grid

Michael Specht and Christine Rosinger

Abstract. The domain of electromobility extends the topic of the Smart Grid by another large consumer, but also adds storage possibilities. Therefore it is necessary to create a new infrastructure of loading points. The communication is not limited between loading points and the vehicle, but there is also a communication between loading point and the distribution network. Additionally there are many different scenarios possible, whereas the loading point also communicates with EMS/DMS. These chapter addresses the aforementioned points and give further information and possible solutions in terms of communication standards

10.1 Introduction

In the last years, electrical vehicles gained more and more importance with the increased interests of politics and the automobile industry. One of the main focuses in the future would be to fully integrate the electrical vehicles into the upcoming Smart Grid. To achieve this it is necessary to work towards an interoperable communication network. One way is to use already established standards of the energy domain. This chapter shows different future scenarios in terms of charging electrical vehicle in the future. At first this chapter introduces necessary evolutionary steps. The subsequent scenarios are build on this evolutionary steps. Finally the possible standards for different parts of the scenario are introduced. At last a conclusion is given.

10.2 Evolutionary Steps

The expected evolutionary steps towards the integration of electrical vehicles into the electrical network can be divided into four major different steps, as shown in figure 10.1.

Michael Specht · Christine Rosinger
OFFIS – Institute for Information Technology, Escherweg 2, 26121 Oldenburg, Germany
e-mail: {michael.specht, christine.rosinger}@offis.de

M. Uslar et al.: *Standardization in Smart Grids*, POWSYS, pp. 163–177.
DOI: 10.1007/978-3-642-34916-4_10

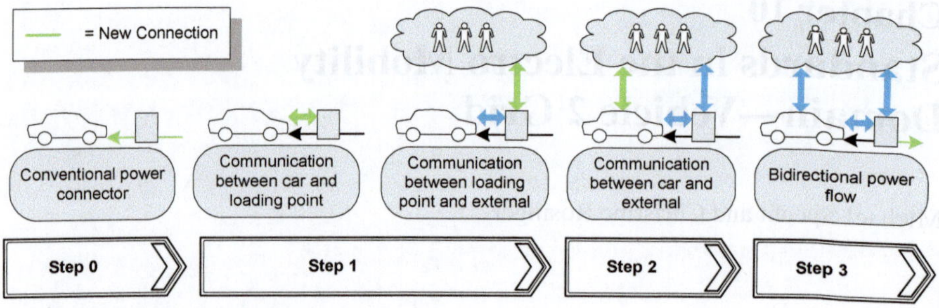

Fig. 10.1 Overview on Evolutionary Steps [14]

The first step **0** describes the status quo. The loading points are from few stable electricity providers or are conventional electric sockets. Payments are only made directly to the service station or the billing is done with existing metering points.

- Step **1** would be an extension of the current loading infrastructure and to made loading points area-wide available. In addition, new billing concepts would be introduced. The charging will autonomically done by the car itself. The requirements for this step includes a communication between the loading station and the car, a clearinghouse for the billing as well as communication technology for the billing process.
- The capability of external control of the charging is the focus of step **2**. This would enable additional features like emergency shutdown or down-regulation of the charging in the case of a grid overload. The external control function could also be used to integrate electrical vehicles or whole loading points into an Virtual Power Plant (VPP). From the technical view, it is necessary to create communication interface for the control options either at the loading point, or in case of mobile telephony systems on board, directly in the vehicle.
- Step **3** deals with the energetic recovery system into the distribution network, which can be used to generate grid services. The flexibility gained through energetic recovery can be used to specific load distribution, balancing power with reserve supply, as well as voltage and frequency maintenance. The technical requirements includes the electronical systems for the energetic recovery as well as the communication interfaces and infrastructure to utilize the new services. A billing system which can be used in both ways, for load and feed-in are also needed.

10.3 Scenarios

The steps mentioned in the previous chapter can be reached in different ways. The following scenarios are used as a basis of assessment for communication standards needed to fulfill the requirements.

10.3.1 Scenario 1: Simple Charging at Non Public Loading Points

Description

Electrical vehicles are charging at conventional electric sockets or special charging sockets. There is no communication between socket and vehicle. The billing takes place with the classical electrical meter and billing system. There are no new challenges for the ICT as shown in figure 10.2. This scenario is available at evolutionary step 0.

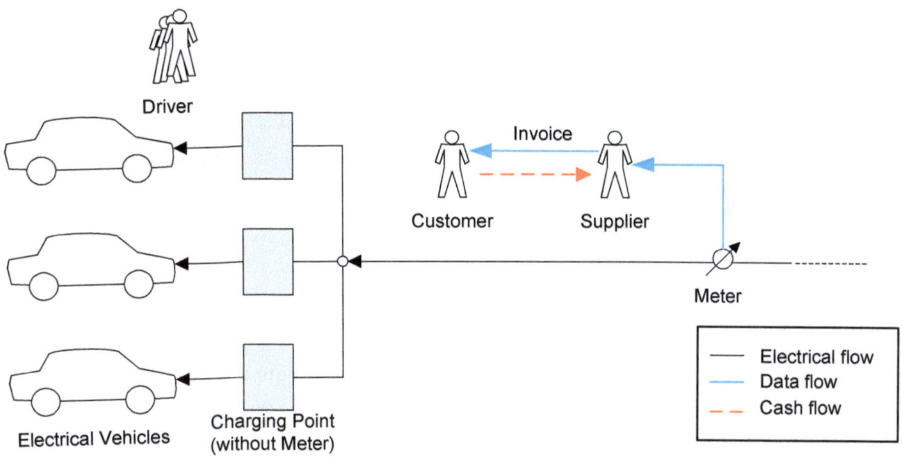

Fig. 10.2 Simple Charging process at non public charging stations (simplified billing) [14]

Business process

The driver connects the vehicle with a conventional plug to a loading point (for example a conventional electric socket at home). The power consumption is detect by the local classic metering device, and invoiced to the consumer related to this metering device. The utilization by third parties is not included in this scenario. The loading and billing processes are separated.

10.3.2 Scenario 2: Simple Charging at Public Loading Points with Direct Payment

Description

The vehicles will be charged at conventional electric sockets or special charging sockets with integrated metering systems. There is no communication needed between vehicle and socket. The billing system will read the charged amount after charging from the metering system and the driver can use established payment methods (similar to gas stations). The charging station operator gets an invoice from the supplier based on the classic centralized metering system. There are no new challenges for the ICT. This scenario is available at evolutionary step 0.

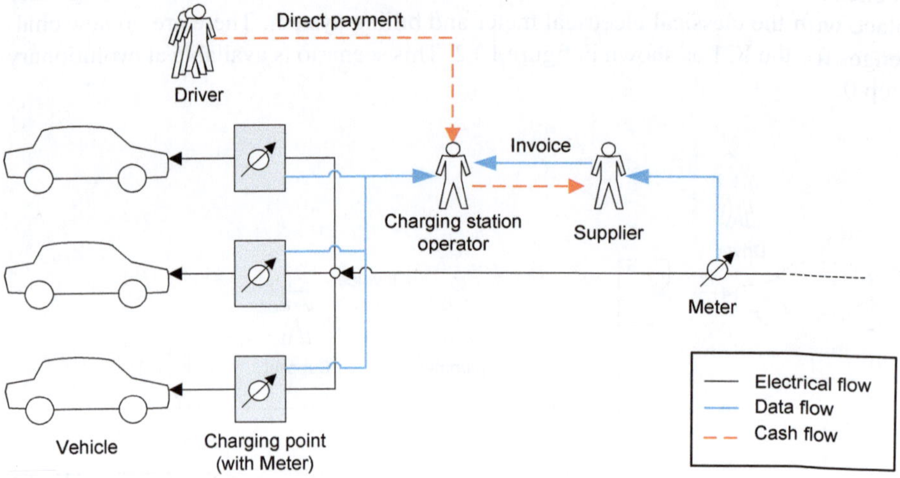

Fig. 10.3 Simple Charging process at public charging stations with direct payment [14]

Business process

The driver connects the vehicle with the charging station. After the charging process (long-term or fast-loading), the driver pays the charging point operator directly based on the metered amount. The charging point operator gets an invoice from the supplier to pay as shown in figure 10.3.

10.3.3 Scenario 3: Battery Exchange at Swap-Out Stations

Description

The battery exchange will be a completely automatized process. The empty battery in the vehicle will be exchanged with a more loaded one. The billing will be based on the distance since the last change (like in the Better Place Project[1]) or based on the consumed energy. The latter case would make a metering device in the battery necessary. This scenario is available at evolutionary step 0.

[1] http://www.betterplace.com/

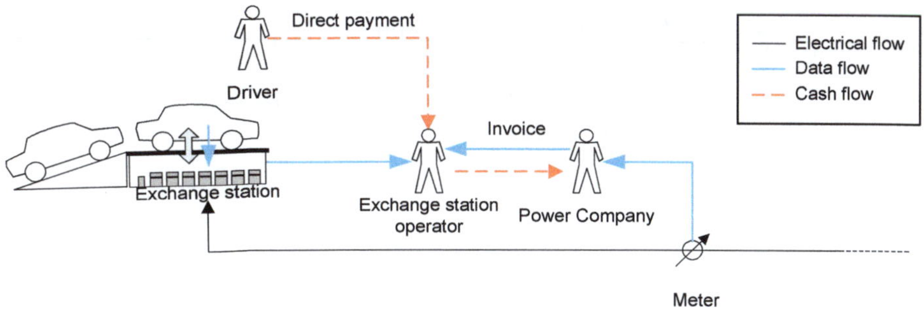

Fig. 10.4 Battery exchange process at an exchange station [14]

Business process

The driver puts the vehicle on/in an exchange station to change the used battery with a more filled one. After charging the power difference will be determined and will be the base for billing. The driver pays directly. The battery will be loaded again for other customer. The exchange station operator gets an invoice from the power supplier to pay as shown in figure 10.4. There are no new challenges from the ICT perspective.

10.3.4 Scenario 4: Roaming with Electricity Provider as Charging Station Operator

Description

This scenario represents a roaming scenario with electricity providers as charging station operators. To charge the electrical vehicle at these charging stations, the owner has to conclude a contract with the electricity provider.

The challenge for the ICT lies in the authentication and the billing in the roaming scenario.

At least evolutionary step 1 is necessary to implement this scenario.

Business process

A customer wants to charge his electrical vehicle at a charging station from his electricity provider (see figure 10.5). The consumed energy will then be charged by the electricity provider, maybe on a monthly basis. The authentication takes place at the charging point, where the currently applicable tariffs are shown.

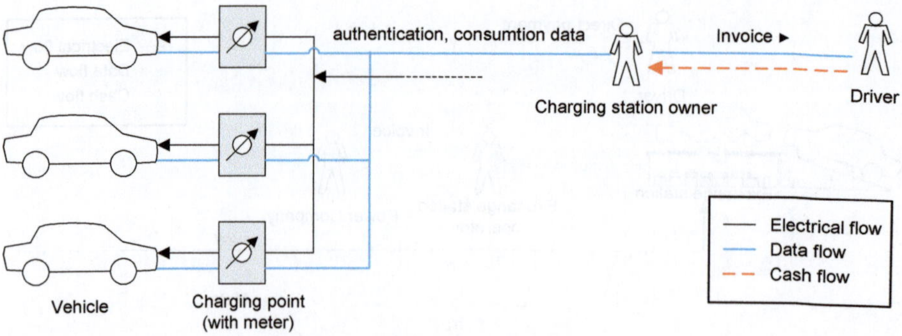

Fig. 10.5 Electricity provider as charging station operator [14]

10.3.5 *Scenario 5: Charging Current Limitation to Prevent Grid Congestion*

Description

This scenario offers the possibility to put a limitation on the charging current to prevent grid congestion. This scenario is limited to necessary intervention at the charging process to prevent emergency shut downs in the network segment. No higher network services are addressed in this scenario. There is no intent to realize a profit out of it, therefore there is no communication for billing as shown in figure 10.6. The possibility to limit the current through charging is generally preferable,

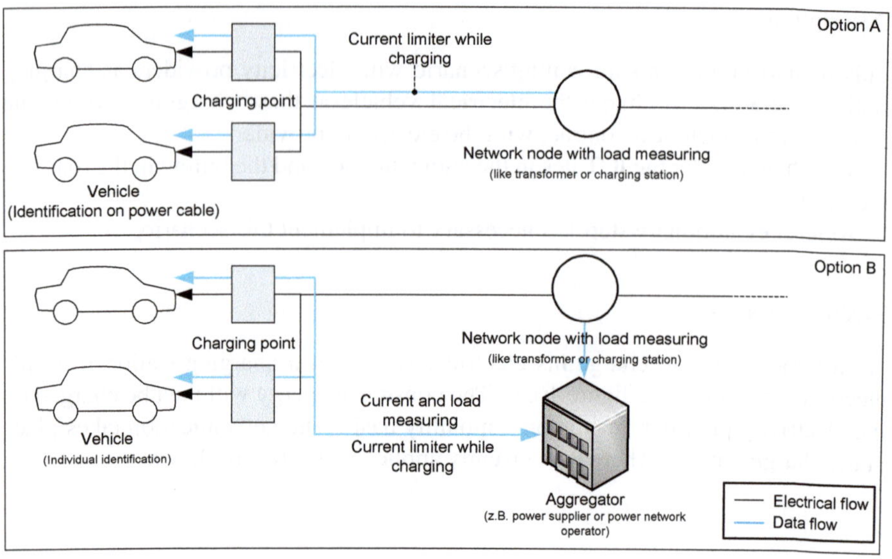

Fig. 10.6 Charging current limitation to prevent grid congestion [14]

especially in locations like parking ramps. To enable this scenario, evolutionary step 2 is necessary.

Option A in figure 10.6 represents an elegant possibility for current limitation with using Powerline Communication(PLC) and without integration additional decision making instances. This easy to implement solution does have the drawback of higher costs for manufacturer independent receiver units in the electrical vehicle or the charging station.

The alternative in option B is to use a centralized network load management. This load management would need models of the network topology as well as measurement data. These would result in a considerably more comprehensive communication infrastructure than option A.

Business process

There is no business process to describe in this scenario.

10.3.6 Scenario 6: Remote Controlled Charging (Demand Side Management)

Description

This scenario deals with the direct control of the charging process, which exceeds the emergency shut downs of the previous scenario 5. This can be used to offer extended services. The motivation to implement such a "Demand Side Management" (DSM) can be splitted into four main points:

1. Network view: Overload management, distribution network operator and charging station operator can lower the charging power to prevent grid overloads (same as in scenario 5)
2. Network view: Offer balancing power or other system services of this kind
3. Power supplier view: react on fluctuating purchasing prices on the market because of loading or weather issues
4. Power plant operator view: A Virtual Power Plant operator can use the offered flexibility to refine the fluctuating power feeding of distributed energy resources(DER)

DSM can either be implemented as a direct control or incentive based. With direct control the decision for the charging process is made at the aggregator (see figure 10.7). With incentiv-based DSM the decision is moved to the vehicle itself, respectively the charging management system.

To directly control the charging process, it is possible to use spontaneous signals as well as schedule based commands for a given time frame.

In case of incentive-based controlling, the easiest way to realize it would be to offer time based tariffs with at least one high and one low cost time frame, for example the next day. However, it seems more effective to send price signals on occasion in more flexible time frames.

To enable this scenario evolutionary step 2 is necessary.

The main challenge is to implement a DSM for electrical vehicles with a homogeneous communication for tariff and control signal exchange between vehicle and service purchaser.

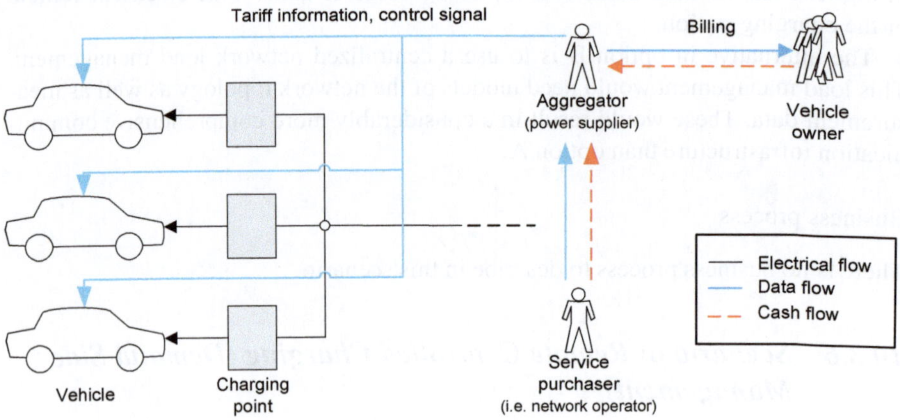

Fig. 10.7 Remote Controlled Charging (Demand Side Management) [14]

Business process

The business process is not described further because of the complexity and variety, so there is no clear business process.

10.3.7 Scenario 7: Charging with Energetic Recovery System

Description

This scenario describes an extended version of scenario 6. In addition to the remote controlled charging, it is possible to add an energetic recovery system. The electric vehicles can be used as energy storages to further enhance the aforementioned grid services as shown in figure 10.8.

Compared to scenario 6, an advanced communication and billing infrastructure is needed to refund the customer as well as limits of the battery.

Evolutionary step 3 is mandatory for this scenario.

Business process

To successfully use this scenario in a reasonable business process a vast amount of vehicles is necessary. Additionally, a the vehicles need to have a communication link to the aggregator, which then has the possibility to manipulate the charging an

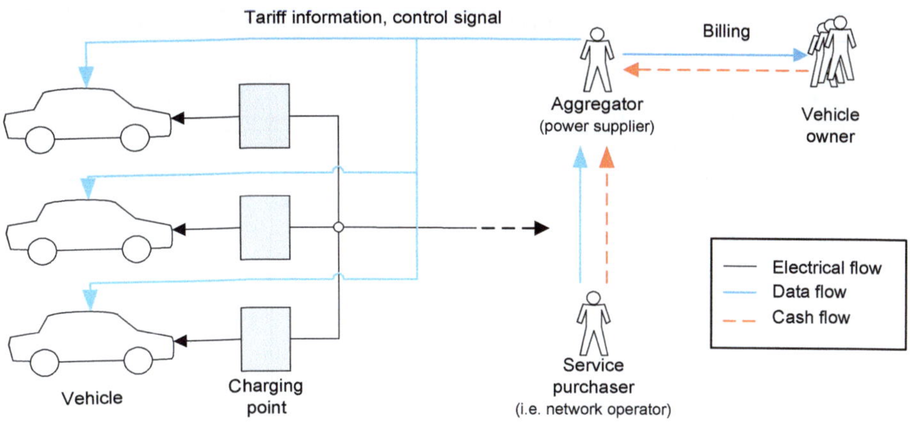

Fig. 10.8 Charging with energetic recovery system [14]

the recovery. This manipulation can either be done with direct command signals or indirectly with tariff informations send to the vehicle.

The further business is not described because of the complexity and variety as already mentioned at scenario 6.

10.3.8 Scenario 8: Metering System Integrated in the Vehicle

Description

If a low dissemination of electrical vehicles is given, a higher amount of charging stations are needed in comparison. To counteract this disadvantage a more adapted charging infrastructure is needed. This scenario describes charging stations as parking site with several charging points. The charging points do not have any electronic system inside. The necessary electronic needed for authentication and billing is only built-in into the electrical vehicle and the centralized charging station.

For this scenario, evolutionary step 1 is needed.

The challenge is to establish a communication link between the charging station and the electrical vehicle for authorization and billing of the charging process.

Business process

The initial state of the charging point is currentless. If an electrical vehicle is connected to the charging point the centralized charging station has to recognize this (either from measures or a switch contact). The charging station initiates a direct communication link to the electrical vehicle and tries to authenticate the vehicle. If this authentication is successful, the charging station enables the charging point (see figure 10.9). The consumption data can either be communicated at the end of the charging process or periodically, and the consumption results in an invoice to the electrical vehicle owner.

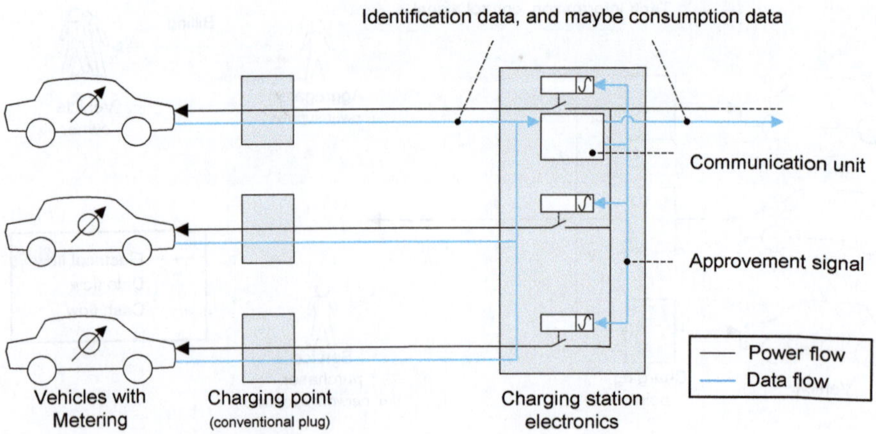

Fig. 10.9 Metering system integrated in the vehicle [14]

10.4 Security for the Electro Mobility Domain

Information security is one of the upcoming Smart Grid issues, which is not considered extensively but becoming more and more important, as has already been motivated in Chapter 8. Different security challenges that are identified in the Smart Grid domain apply to the electro mobility subdomain as well. Beyond the functional safety of vehicles, which can also be compromised with the exploitation vehicle security issues, a more system relevant scenario is to be considered when vehicles are attached to the grid. Electric vehicles (or more specifically their batteries) can, beyond charging, be used for sophisticated Smart Grid applications like load balancing. With an increasing number of electric vehicles and relying on such applications, electric vehicles can become also grid-critical elements. Hence, the communication of electric vehicles with the Smart Grid connection point (e.g., for charging purposes) shall be analyzed and assessed regarding security-relevant issues. This comprises for example billing and privacy aspects. To conduct a holistic security analysis, a continuous security assessment during all evolutionary steps in the context of the vehicle to grid implementation (as described in Section 10.2) is required.

10.4.1 Security Scenario for the Electro Mobility Domain

Typically security issues are identified according to specific scenarios which can be analyzed in detail. Figure 10.10 outlines a simple, generic scenario in the context of electro mobility, covering basic aspects of vehicle to grid communication. On this basis, of course further scenarios are to be elaborated covering the aspects in more detail.

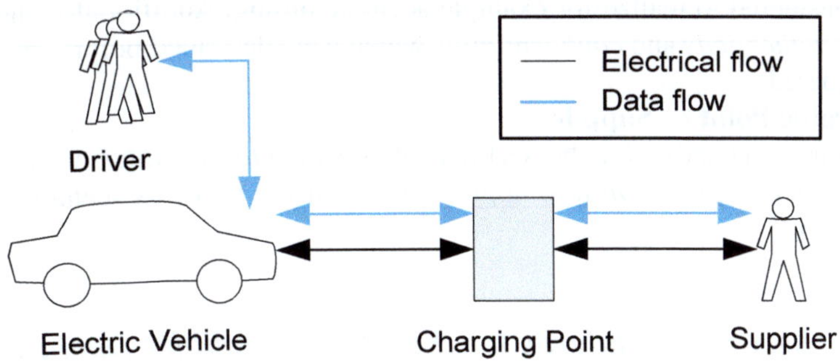

Fig. 10.10 A generic scenario for electric vehicles

In this depicted scenario, different actors are involved to exchange power *(power flow)* and data *(data flow)*. The *electric vehicle* communicates with the *driver*, for example using a display where information is shown. The driver itself can for instance use a keyboard to enter information in order to communicate with the vehicle. On the one hand, the *electric vehicle* is supplied with electricity over the *charging point* which is operated by the *supplier*. On the other hand, data is exchanged between these three actors, e.g., identification data, consumption data, or information regarding the battery state. This scenario identifies multiple interfaces between the actors where diverse threat scenarios may be applied. Basically, all these data interfaces and also the actors themselves must be included in security considerations and assessments.

In the following itemization lists some exemplarily threats for the different interfaces and actors, providing a basic security assessment.

- **Electric Vehicle**
  The electric vehicle needs a strong *authentication* measure to avoid that an attacker can use the authenticity of the vehicle for another vehicle to manipulate for example the billing process. Also strong measures for confidential data are necessary because otherwise tracking profiles could be created by unauthorized persons.
- **Driver ↔ Electric Vehicle**
  For the communication between the driver and the electric vehicle also *authenticity* measures are required. In general, the actual driver of the vehicle shall be charged for the obtained electricity. Thus, the driver has to use his own authentication for the car, which in addition avoids that, in case of theft, the thief can recharge the vehicle.
- **Electric Vehicle ↔ Charging Point**
  For the communication between the electric vehicle and the charging point the security goal *non-repudiation* is important. Neither the electric vehicle nor the charging point shall revoke the charging. Also the security goal *integrity* must

be considered to realize for example accurate billing. Additionally, the security goals *authenticity* and *confidentiality*, which were described before, shall also be considered.

- **Charging Point ↔ Supplier**
  The threat scenarios for the security goals *authenticity*, *integrity*, *confidentiality* and *non-repudiation* also apply—just as in the previous scenarios—to this scenario.

10.4.2 Security Standards for the Electro Mobility Domain

Up to now, there have no specific security standards for the security domain been released yet. However, the security standards described in Section 8.3, can also be used for securing electric vehicles in the Smart Grid. For this purpose these existing standards have to be analyzed, reengineered, or new standards will even have to be developed. Currently, there is only one security standard under development, named ISO/IEC 15118 and entitled as "Road vehicles – Vehicle to grid communication interface". This standard under development, especially considers electric vehicles and their communication. More precisely, ISO/IEC 15118 specifies the communication, on the one hand, between electric vehicles and charging stations and, on the other hand, between charging stations and back-end infrastructure. This standard consists of the following three parts and to be applied in the context of electric vehicles after its release.

- Part one describes general information for this subdomain and provides some use case definitions as well as it illustrates related requirements.
- Part two gives a technical protocol description and specifies the requirements of the open systems interconnections (OSI).
- Part three examines the physical layer and data link layer requirements for this subdomain.

Another standard, the "Protection Profile for Smart Metering" (described in Section 8.3.6), is a possible candidate for the protection of metering data for electric vehicles. After this outline of security specific standards, a broader view on applicable ICT standards shall be given in the next section.

10.5 Existing Standards Relevant for ICT in the Electro Mobility Domain

The momentarily existing and available standards (shown in figure 10.11) in the energy domain can already be used for the integration of electrical vehicles into the Smart Grid regarding ICT. Many of these standards are especially made for the future Smart Grid and therefore, the electromobility is already integrated. The electric vehicles themselves are not in the focus of these standards but the communication process of integrating them. The integration of the electromobility relevant parts into the standards of the electricity domain shows how much the electricity domain is involved.

Bereich	Standard oder Norm	Titel
Electricity industry	IEC 60870	Telecontrol equipment and systems
	IEC 60870-5	Part 5: Transmission protocols
	IEC 60870-6	Part 6: Telecontrol protocols compatible with ISO standards and ITU-T recommendations
	IEC 61968 & IEC 61970	Application integration at electric utilities – System interfaces for distribution management & Energy management system application program interface (EMS-API)
	IEC 61850	Communication networks and systems in substations
	IEC 61850-7-410	Part 7-410: Hydroelectric power plants - Communication for monitoring and control
	IEC 61850-7-420	Part 7-420: Basic communication structure - Distributed energy resources logical nodes
	IEC 62056	Electricity metering – Data exchange for meter reading, tariff and load control
	IEC 62056-21	Part 21: Direct local data exchange
	IEC 62056-62	Part 62: Interface Classes
	IEC 62325	Framework for energy market communications
	IEC 62357	Power system control and associated communications - Reference architecture for object models, services and protocols
	IEC 61140	Protection against electric shock – Common aspects for installation and equipment
	IEC 62040	Uninterruptible power systems (UPS)
	IEC 60529	Degrees of protection provided by enclosures (IP Code)
Electromobility	IEC 61851	Electric vehicle conductive charging system
	IEC 61439-5	Low-voltage switchgear and controlgear assemblies - Part 5: Assemblies for power distribution in public networks
	IEC 62196-1	Plugs, socket-outlets, vehicle couplers and vehicle inlets – Conductive charging of electric vehicles - Part 1: Charging of electric vehicles up to 250 A a.c. and 400 A d.c.
	IEC 62196-2	Plugs, socket-outlets, vehicle couplers and vehicle inlets - Conductive charging of electric vehicles - Part 2: Dimensional interchangeability requirements for a.c. pin and contact-tube accessories
	IEC 62196-3	Plugs, socket-outlets, and vehicle couplers - conductive charging of electric vehicles - Part 3: Dimensional interchangeability requirements for pin and contact-tube coupler with rated operating voltage up to 1 000 V d.c. and rated current up to 400 A for dedicated d.c. charging
	ISO/IEC 15118	Work in Progress, Road vehicles – Vehicle to grid communication interface
	ISO TC22/SC3 JWG V2G	Vehicle to grid communication interface (V2G CI - Kommunikationsschnittstelle)
	ISO 6469-3	Work in Progress, Electric propelled road vehicles – Safety specifications – Part 3 Protection of persons against electric shock
Automobile industry	ISO 11898	Road vehicles - Controller Area Network (CAN)
	ISO 10681	Communication on FlexRay
	MOST	Media Oriented Systems Transport
	De-facto-Standard	Local Interconnection Network (LIN)
	AUTOSAR	AUTomotive Open System ARchitecture
	FIBEX	Field Bus Exchange Format

Fig. 10.11 Existing standards relevant for ICT in the electro mobility domain [14]

The mapping of the standards itself can be divided into different groups like billing, network status monitoring, control, and tariff information.

The billing group includes the standards for the measurement of consumption data—which includes the Smart Message Language (SML) [2], IEC 62056 DLMS/COSEM [4, 3, 9, 10, 8, 7, 12], ANSI C12.19 [1] and Metering Bus (M-Bus) [15]—as well as the extended communication for the consumption data. It may be necessary to include more information additional to the main consumption data, which then can be handled with the standards IEC 61850 [5] and IEC 61968/61970 [11, 6].

The network status monitoring is mentioned in scenario 5. If these measured values have to be communicated, suitable data have to be used. In this case, the IEC 61850 [5] and IEC 61968/61970 [11, 6] can be used.

The IEC 61850 [5] and IEC 61851 [13] are suitable to be used in control based scenarios to coordinate electricity recovery and charging processes.

The indirect control of the charging process with the help of tariffs can be handled with the already mentioned IEC 61850 [5] and IEC 61968/61970 [11, 6].

10.6 Conclusion and Outlook

Todays available standards and techniques for ICT are basically sufficient for the integration of the mentioned electromobility scenarios. Although the standards offer good approaches, many problems are remaining. Beside the missing loading infrastructure a missing comprehensive data network are restricting the successful integration of electrical vehicles.

The inconsistent data models are another gap. The existing standards for communication in the energy domain, IEC 61850 and IEC 61968/61970, offer the possibility to extend the data model, but these extensions have to be based very close on the business process to be successful. Another important point at extending the data models is to consider other business process in the energy domain—for example the emergency shut down or charging current limitations can also be used on other Smart Grid components like Combined Heat and Power plants (CHP)—to avoid further problems.

References

1. AEIC: AMI Interoperability Standard Guidelines for Communications and Supporting Enterprise Devices, Networks and Related Accessories (2010)
2. EMSYCON: Smart Message Language Version 1.03 (2008)
3. IEC: 62056-21 ed1.0: Electricity metering - Data exchange for meter reading, tariff and load control - Part 21: Direct local data exchange (2002)
4. IEC: 62056-42 ed1.0: Electricity metering - Data exchange for meter reading, tariff and load control - Part 42: Physical layer services and procedures for connection-oriented asynchronous data exchange (2002)

5. IEC: 61850-1 ed1.0: Communication networks and systems in substations - Part 1: Introduction and overview (2003)
6. IEC: 61970-1 Ed.1: Energy management system application program interface (EMS-API) - Part 1: Guidelines and general requirements (2005)
7. IEC: 62056-47 Electricity metering - Data exchange for meter reading, tariff and load control - Part 47: COSEM transport layers for IPv4 networks (2006)
8. IEC: 62056-53 ed2.0: Electricity metering - Data exchange for meter reading, tariff and load control - Part 53: COSEM application layer (2006)
9. IEC: 62056-61 ed2.0: Electricity metering - Data exchange for meter reading, tariff and load control - Part 61: Object identification system, OBIS (2006)
10. IEC: 62056-62 ed2.0: Electricity metering - Data exchange for meter reading, tariff and load control - Part 62: Interface classes (2006)
11. IEC: 61968-1: Application integration at electric utilities - System interfaces for distribution management - Part 1: Interface architecture and general requirements (2007)
12. IEC: 62056-46 Electricity metering - Data exchange for meter reading, tariff and load control - Part 46: Data link layer using HDLC protocol (2007)
13. IEC: 61851-1 Edition 2.0 Electric vehicle conductive charging system Part 1: General requirements (2010)
14. Mayer, C., Tröschel, M., Uslar, M.: Elektromobilität: Geschäftsmodelle, Kommunikation und Steuerung (2012)
15. Universität-GH Paderborn: M-Bus (1997)

Chapter 11
Smart Metering in the European Context

Michael Specht

Abstract. Smart Meters can be a major part of the future Smart Grid, as the are seen as a key technology to connect customers and enable the participation of the customer in the Smart Grid. Further a wide range of extended Smart Grid functionalities such as Demand and Response are supported. This chapter introduces the report of the Smart Meter Coordination Group to the European Smart Meter Mandate M/441. Furthermore, exemplary standards out of this report are described. In addition, several standards are presented, which are not included in the above mentioned report, but relevant to the standardization community.

11.1 Introduction

The Smart Metering domain is seen as one of the main enablers in the Smart Grid by several experts mentioned in different international roadmaps like [20, 18, 4]. However no central definition of Smart Grid functionalities exist, but often includes different domains such as metering and home automation. Smart Meter describes usually a metering system for electrical power (but others are possible as well like gas, heat, etc.) which records and communicates the data to different systems like Energy Management Systems or external systems for billing, control, etc. These functionalities enable or support different Smart Grid services like Demand Side Management (DSM), integration of Distributed Energy Resources (DER), pricing signals, interfaces to home automation and utility companies, and of course detailed output of the energy consumption on e.g. home displays. Hence Smart Meter can increase the self awareness of the customer it can lead to energy savings but also supports efficient usage of house appliances.

Michael Specht
OFFIS – Institute for Information Technology, Escherweg 2, 26121 Oldenburg, Germany
e-mail: `michael.specht@offis.de`

M. Uslar et al.: *Standardization in Smart Grids*, POWSYS, pp. 179–188.
DOI: 10.1007/978-3-642-34916-4_11

Therefore, the European Union (EU) promotes the introduction of Smart Metering systems through different recent legislations to support the Smart Grid deployment. This should lead to a large-scale rollout of the electricity and gas meters with additional functionalities in Europe. Greater energy efficiency awareness by end users as well as the resulting potential for energy savings is the main driver for this initiative.

One of the main problems is not the metering itself, but the outbound communication links to, e.g., metering aggregation servers or inwards to systems such as inhouse automation system.

This chapter addresses mainly the problem of the interoperable communication and introduces some standards and initiatives, which provide solutions for those problems. Section 11.2 will outline the mandate M/441 and the efforts of the Smart Meters Coordination Group regarding this matter. The subsequent sections will describe different standards used in the Smart Meter domain. Finally this chapter concludes with a summary in Section 11.10.

11.2 CEN/CENELEC/ETSI Smart Meters Coordination Group Report for EU-Mandate M/441

One of the main actions of the EU was to issue the Mandate M/441 [8] to CEN, CENELEC, ETSI to standardize Smart Meter functionalities and communication interfaces. The results of this mandate are standards and technical documents. Standards are freely usable technical specifications and technical rules for products and systems. The main objectives are to ensure interoperability, customer protection, and system reliability. Therefore, six main aspects of Smart Metering are examined:

- Reading and transmission of metering data remotely
- Two way communication between meter and market participant (biller)
- Support of different tariff models and payment systems by the metering system
- Remote shutdown of the meter and the possibility to start/stop the supply
- Communication with devices inside the household
- Support a display or an interface in the household to show metering data in real time

Existing standards are classified by the Smart Meters Coordination Group (SM-CG) into the aforementioned six functionalities and the responsibilities are delegated to individual standardization organizations as shown in Figure 11.1. The same figure shows the relevant communication hubs. The *Central communication system* is the "communication head" of a Smart Metering system. The data is sent by meters or meter data concentrators through public or private Wide Area Networks (WAN). The system which uses such a central system is a *Smart Meter (M2M) gateway*, which can either be equipped inside any Smart Meter or be integrated as a separate system. This gateway represents the entry point to the house. Interactors of M2M remote gateways are *Electricity meters*, *Non-electricity meters* (generally battery powered) or *Home automatiom* and customer information systems (e.g. displaying the current energy consumption). The lower part of Figure 11.1 shows the external connections which are also impacted by Smart Metering by using the gathered data or sending signals for further Smart Grid functionalities described earlier.

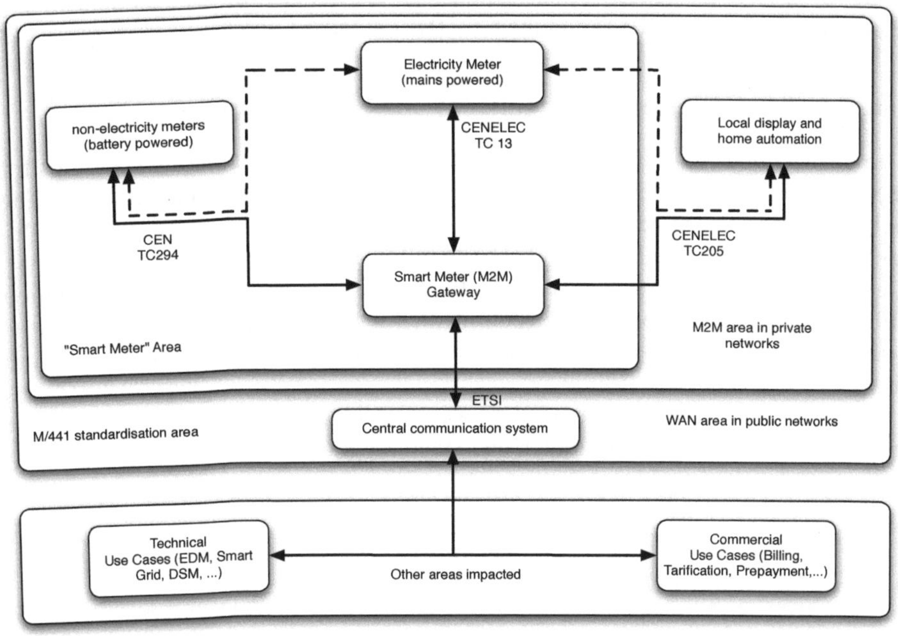

Fig. 11.1 SM-CG standardization organization reference architecture [4]

The following paragraphs introduce interfaces for the different communication paths related to Smart Meters and an overview on existing relevant standards.

Interface Between Smart Meter and Smart Meter Gateway

This communication path is necessary if the metering device is a stand-alone physical unit and shall ensure the interoperability between the metering device and other systems like between gateway/electricity meters and battery-powered meters, gateway/electricity meters and the home automation system, and gateway/electricity meters and concentrator/central communication systems.

The standard recommended for this communication path is the IEC 62056 "Device Language Message Specification (DLMS)/Companion Specification for Energy Metering (COSEM)". An exception for this recommendation are battery powered metering devices. Because of the limited amount of power and the designated life time (typically for at least 10 years), a more battery power saving standard is recommended: the EN 13757 M-Bus standard.

Interface Between Home automation and Smart Meter Gateway

To support the extended functionalities in the home automation an EMS, these systems need access to data on energy consumption or further parameters from Smart Meter devices. Existing European standards are the EN 50090 "Home and Building Electronic Systems (HBES)" and the EN 50491 standard series.

Interface Between Smart Meter Gateway and External Systems

The SM-CG does not offer specialized recommendations on existing standards for these interfaces, but identifies necessary standardization work on this part. The only standards recommended are communication standards like the public cellular mobile network standards (Global System for Mobile Communications (GSM), General Packet Radio Service (GPRS), Enhanced Data Rates for GSM Evolution (EDGE), and Universal Mobile Telecommunications System (UMTS)) and 3rd Generation Partnership Project (3GPP) Standards like Long Term Evolution (LTE) and more.

To further extend the recommendations, the following sections will describe the most relevant of the already mentioned standards in detail, but will also provide recommendations on additional standards.

11.3 IEC 62056 DLMS/COSEM

DLMS [13, 10, 14, 11, 12, 9, 15] is an international standard series from the IEC, which is used to request data from Smart Meters at the consumer site. The specification is developed by an international company consortium with over 60 members, the IEC, and CEN. DLMS/COSEM is chosen as one of the main standards in the Smart Metering domain by the SM-CG report.

DLMS defines different transport protocols as well as communication objects for energy, gas, water, and heat metering devices. The DLMS/COSEM specification (described in [5]) follows a three step approach as shown in Figure 11.2.

The *first step* is about modeling metering equipment and is called "Modelling" accordingly. This includes rules for data identification as well. Additionally the data model provides a view of the functionality of the meter through generic building blocks. However the model does not cover internal, implementation-specific issues.

The *second step*, which is called "Messaging", covers the communication services and protocols for mapping the elements of the data model to Application Protocol Data Units (APDU).

The *last step* named "Transporting", addresses the services and protocols for the transportation of the messages through communication channels.

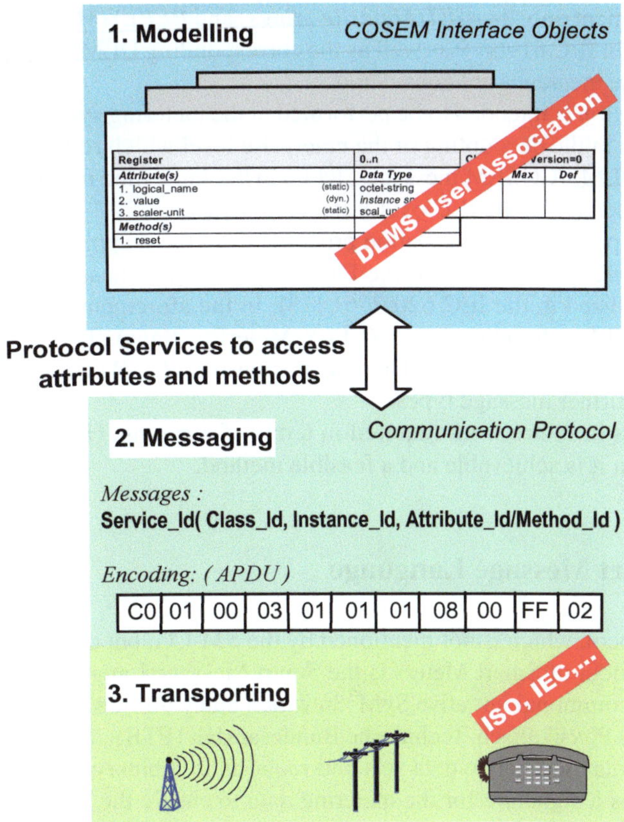

Fig. 11.2 DLMS/COSEM three step approach [5]

In summary, IEC 62056 DLMS/COSEM is a complex standard series with many facets. A starting point for further information would be either the standard documents themselves or the DLMS/COSEM community website[1].

11.4 Harmonization of DLMS and CIM

As already mentioned before in Section 11.2, the Smart Meter devices are connected to external systems, which again are parts of systems like SCADA or billing systems. Chapter 6 has described the Common Information Model (CIM) which is one of the essential upcoming standards in this application area. To facilitate the interoperability between DLMS and CIM the technical report "Translation between CIM and DLMS/COSEM message profiles" [16] was written.

[1] `http://www.dlms.com`

At this moment only the *GET(MeterReadings)* and *REPLY(MeterReadings)* messages defined in IEC 61968-9 as well as the corresponding DLMS GET-Request and GET-Response messages are considered.

The translation is intended to be performed in the metering system head-end. The advantage lies in the decoupling of the enterprise level which reduces the burden of interpreting DLMS/COSEM messages, while in the same time maintaining standard interfaces as defined in the IEC 61968 standard series.

The major point for the translation process lies in the mapping between the IEC 61968 "ReadingType" codes and the DLMS Object Identification System (OBIS) codes standardized in the IEC 62056-61 [13]. In the aforementioned case of mapping the GET(MeterReadings) and REPLY(MeterReadings), the mapping is found to be relatively easy to be mapped and the responsible IEC Working Group expects the same for further message types.

This results in future work required in terms of mapping. However, the first results show that it is achievable and a feasible method.

11.5 Smart Message Language

Another protocol, which is not mentioned by the SM-CG, but can be used for local or remote readout of Smart Meters is the Smart Message Language (SML) [19]. It is part of the communal initiative SyM^2 in which many utilities, metering manufacturers, and the Physikalisch-Technische Bundesanstalt (PTB) are part. The protocol specifies a syntax to encode data sets and requests to a binary format. Part of this specification is a signature for the metering data to ensure the integrity of the data later on. SML does not specify its own transport protocol, but offers the opportunity to use different communication protocols, for example: serial connections or standardized network protocols. The first successful field experiments with compatible devices were made in 2009 [6].

11.6 Metering Bus

The Metering Bus (M-Bus) is an European standard mentioned in the SM-CG report for the readout of battery powered Smart Meter devices under the EN 13757-3 standard [3]. The standard was developed as a joint work of the University of Paderborn as well as the companies Techem and Texas Instruments.

M-Bus defines a bus system with serial data transfer on a duplex wire. Power supply for sensors using this wire is possible but limited. The bus system supports up to 250 devices. This simple structure allows an affordable implementation into devices with a low power consumption. Meanwhile the standard has been extended with data transmission using short-distance radio on the 900 MHz range.

The defined application layer in EN 13757-3 is basically usable on other transport protocols, not mentioned in the standard. However, exchangeability cannot be guaranteed. Because of insufficient specifications it is necessary to adapt the reader device specifically to the meter.

11.7 ANSI C12

Instead of the IEC 62056 standard series, mentioned in the SM-CG report, in North America ANSI C12 standards are used for metering protocols [7]. The ANSI C12.19 standard presents common structures for:

- Encoding data in communication between end devices (meters, home appliances, ANSI C12.22 nodes)
- Utility enterprise collection and control systems

For these purposes the standard uses binary codes and XML content. The integrated data tables support gas, water, and electricity-sensors-related appliances. In 2008 a major revision of the standard was made which resulted in new data tables, XML-based table description language (TDL/EDL), and the documentation of services and behaviors. It now features new and updated procedures, controls, and definitions. The in 2010 announced revision should include defining the common meter data tables that are required to enable Smart Grid applications such as Demand Response (DR) and real time usage information. The Exchange Data Language (EDL), which is a part of the ANSI C12.19, can be used to constrain new features into a well known structure.

Historically, the ANSI C12.19-1997 [1] was developed in a joint effort by ANSI C12, Industry Canada, and IEEE SCC31 in order to be the leading North American standard for metering data description.

Another subpart of the ANSI C12 standard, C12.22 [2], provides a common application layer for Smart Meters. Also included is a description of the process of transporting ANSI C12.19 table data over different networks. Thereby, ANSI C12.22 supports both, the sessionless communication as well as supporting session communication. For the security part it also provides mechanism to use Advanced Encryption Standard (AES) encryption methods.

11.8 KNX

KNX, standardized as ISO/IEC 14543-3, is not directly connected or specialized in Smart Meter communication, but is a well known home automation standard [17]. Furthermore, it is mentioned in the SM-CG report to M/441. The reason for mentioning a home automation standard is because of the connection required for home automation servers to the metering devices. Without these connections the systems cannot offer the full range of services expected from a home automation EMS.

The standard itself is a follow-up of the three bus standards European Installation Bus (EIB), BatiBUS, and European Home Systems (EHS) and is mostly identical with the last released version of EIB.

KNX usually utilizes serial communication on two-wire lines, which can be used to deliver power to sensors or actors. KNX-RF is an extension of the KNX standard, which defines a wireless connection using short-distance radio with 868 Mhz. It can be used to connect wireless sensors and actors within a range of 10 meters.

KNX is the leading home automation standard in Europe and uses the same physical transmission layers like "Wireless M-Bus", previously described in Section 11.6

11.9 ZigBee Smart Energy Profile

ZigBee is a specification for communication protocols using the IEEE 802 standards family [21]. It is only mentioned briefly in the the SM-CG report, but can be seen as a further recommendation. ZigBee "Smart Energy" is a subpart and profile with specialized features for the upcoming Smart Grid and home automation domain.

In general, ZigBee is a low-cost, low-power wireless mesh network standard developed by the ZigBee alliance. The ZigBee alliance is a group of companies and organizations counting over 400 members. Besides, the "Smart Energy" which is the most interesting part for this domain many other parts have been developed as be named in the following list:

- ZigBee Home Automation
- ZigBee Telecom Service
- ZigBee Health Care
- ZigBee Remote Control
- ZigBee Input Devices

The Smart Energy standard was developed especially considering Smart Grid requirements. The latest version, Smart Energy Profile (SEP) 1.2, covers already a great part of features for the Smart Grid, namely:

- Basic metering (measurements, historical, etc.)
- Demand Response and Load Control
- Pricing (multiple units and currencies as well as price tiers, etc.)
- Text messages
- Device support for Programmable Communicating Thermostats, Load Controllers, Energy Management Systems, and Home Displays
- Security which allows role-based access like consumer only, utility only, or shared networks
- Multi-utility support for energy, water and gas

Over the last years, a new major revision of the Smart Energy is under development and will be called Smart Energy Profile 2.0 [22]. The focus lies on further extending the abilities for the future Smart Grid as well as security extensions, which are already a major part of the previous version. The extensions that are intended to be supported in the near future are:

- Plug-in electric vehicle charging
- Installation configuration and firmware download
- Pre-pay services
- User information and messaging
- Common Information Model (CIM) support

Especially the last point, the CIM (see Chapter 6) support can be a main advantage in the future, because external systems can be operated with it and there will be less demand to integrate the various data out in the several systems in the Smart Grid.

Table 11.1 Smart Meter standards overview

Standard	Main Application	Considered by the SM-CG
IEC 62056 DLMS/COSEM	Smart Meter Communication	yes
Common Information Model (CIM)	Utility Application	partly
Smart Message Language (SML)	Smart Meter Communication	no
M-Bus	Smart Meter Communication	yes
ANSI C12	Smart Meter Communication	no
KNX	Home Automation	yes
Zigbee Smart Energy	Home Automation	partly

11.10 Conclusion and Outlook

Smart Meters are one of the main pillar of the future Smart Grid as they benefit Smart Grid applications such as DR/DC, tariffing or V2G. They serve as focal point for data access and control in the customer premises domain, but also in the industrial context. To make use of the massive information, communication is needed. This chapter described quite a few possibilities of using standards for inhouse communication, but also with external entities and Table 11.1 shows an overview on the mentioned standards in this chapter.

Unfortunately, the diversity of standards is one of the problems in the Smart Meter domain. Interoperability is needed to easily integrate the inhouse sensors. One of the future tasks in this domain should be to select a main standard and to develop interfaces to others in an interoperable manner. This becomes especially relevant in liberalized markets, to enable exchangeability of systems and devices.

Another important factor for the functionalities and the acceptance of Smart Metering is security. The security and of course privacy issues are not only restricted to Smart Metering, but most of the use cases in the Smart Grid are concerned. Chapter 8 gives an detailed overview on the security aspects in the Smart Grid and the standards in this domain.

References

1. ANSI: C12.19 For Utility Industry End Device - Data Tables (2009)
2. ANSI: C12.22 Protocol Specification For Interfacing to Data Communication Networks (2009)
3. DIN: EN 13757-3: Kommunikationssysteme für Zähler und deren Fernablesung - Teil 3: Spezielle Anwendungsschicht (2011)
4. DKE: The German Standardization Roadmap E-Energy/Smart Grid. VDE (2010)
5. DLMS User Association: COSEM Identification System and Interface Classes (2010)
6. EMSYCON: Smart Message Language Version 1.03 (2008)
7. EPRI: The Electrinet: A Communications Architecture for a Competitive Electric Power Industry (2004)
8. European Commission: M/441 Standardisation mandate to CEN, CENELEC and ETSI in the filed of measuring instruments for the development of an open architecture for utility meters involving communication protocols enabling interoperability (2009)
9. IEC: 62056-21 ed1.0: Electricity metering - Data exchange for meter reading, tariff and load control - Part 21: Direct local data exchange (2002)
10. IEC: 62056-42 ed1.0: Electricity metering - Data exchange for meter reading, tariff and load control - Part 42: Physical layer services and procedures for connection-oriented asynchronous data exchange (2002)
11. IEC: 62056-47 Electricity metering - Data exchange for meter reading, tariff and load control - Part 47: COSEM transport layers for IPv4 networks (2006)
12. IEC: 62056-53 ed2.0: Electricity metering - Data exchange for meter reading, tariff and load control - Part 53: COSEM application layer (2006)
13. IEC: 62056-61 ed2.0: Electricity metering - Data exchange for meter reading, tariff and load control - Part 61: Object identification system, OBIS (2006)
14. IEC: 62056-62 ed2.0: Electricity metering - Data exchange for meter reading, tariff and load control - Part 62: Interface classes (2006)
15. IEC: 62056-46 Electricity metering - Data exchange for meter reading, tariff and load control - Part 46: Data link layer using HDLC protocol (2007)
16. IEC: Translation between CIM and DLMS/COSEM message profiles (2011)
17. KNX Association: KNX System Specifications (2009)
18. NIST: NIST Framework and Roadmap for Smart Grid Interoperability Standards (2010)
19. OFFIS, SCC Consulting, management coaching, M.: Untersuchung des Normungsumfeldes zum BMWi-Förderschwerpunkt 'E-Energy - IKT-basiertes Energiesystem der Zukunft' (2009)
20. SMB Smart Grid Strategic Group (SG3): IEC Smart Grid Standardization Roadmap (2010)
21. ZigBee Alliance: ZigBee Smart Energy Profile 1.2 (2008)
22. ZigBee Alliance: ZigBee Smart Energy Profile 2.0 (2009)

Part IV
Future Applications and Outlook

Chapter 12
OPC UA: An Automation Standard for Future Smart Grids

Sebastian Rohjans and Michael Specht

Abstract. In this chapter, the OPC Unified Architecture is introduced as a future key technology for realizing a variety of Smart Grid applications addressing automation and control. OPC UA is the successor of the established Classic OPC specifications and state of the art regarding information exchange in the industrial automation branch. One of its major improvements is that the application area is no longer limited to industrial automation. Thus, OPC UA can be applied almost in every domain facing challenges in automated control. Besides communication services, information modeling is the key concern of OPC UA. For adopting OPC UA in the context of Smart Grids, three important data structures—the CIM, the IEC 61850, and IEC 61131-3 for industrial control programming—have been identified to be integrated into OPC UA communication. Within this chapter, the OPC UA is basically introduced and its historical development is described. Furthermore, underlying principles for information modeling and communication services are explained. After taking profiling and security concepts of the OPC UA into consideration, the mappings to Smart Grid semantics are analyzed in a more detailed way.

12.1 Introduction and History

Years ago, in the field of industrial automation similar changes compared to the energy domain took place. ICT-applications were implemented in order to support the production processes. However, in the first step these applications were monolithic and have now been replaced by reusable software components. As a result, performant Supervisory Control and Data Acquisition (SCADA)- and Human Machine

---

Sebastian Rohjans · Michael Specht
OFFIS – Institute for Information Technology, Escherweg 2, 26121 Oldenburg, Germany
e-mail: {sebastian.rohjans,michael.specht}@offis.de

M. Uslar et al.: *Standardization in Smart Grids*, POWSYS, pp. 191–209.
DOI: 10.1007/978-3-642-34916-4_12

Interface (HMI)-systems with vendor-specific proprietary interfaces were developed. This induced a very low level of interoperability for the application layer. To oppose this development, an initiative called OPC Task Force was founded. Later, the initiative was renamed to OPC Foundation[1] and aimed at the realization of a standard, which enables real-time data access based on Object Linking and Embedding/Distributed Component Object Model (OLE/DCOM)-technologies for Windows machines.

The OPC Unified Architecture (UA) is a (relatively) new series of standards with its roots in the field of industrial automation [10]. Since years, the Classic OPC standards are dominating this area. The OPC UA is introduced as the successor of the established OPC specifications, i.e. Classic OPC. OPC UA is the state of the art concerning information exchange in the industrial automation branch. Due to its improvements, the application area is no longer limited to industrial automation but OPC UA can be applied almost in every domain. This is due to the underlying generic and object-oriented approach of the OPC UA [9].

One domain, in which OPC UA can be used for information exchange aiming at control, monitoring, and automation of devices and systems is Smart Grids [8]. For the adoption of OPC UA for Smart Grids, the two most important data models (Common Information Model (CIM) and IEC 61850) were identified to be integrated by OPC UA communication [7, 25]. Furthermore, IEC 61131-3 is an essential standard to be integrated with OPC UA. These data models provide domain-specific information required to apply the OPC UA in the energy domain.

The Classic OPC standards were the first results of the OPC Foundation. Today, these standards are implemented for almost all systems within the industrial automation. Briefly, they deal with reading, writing, and monitoring of process data (OPC Data Access (DA)), sending messages in consequence of certain events and alarms (OPC Alarms & Events (AE)), and accessing historical data (OPC Historical Data Access (HDA)). Based on these basic standards, extending specifications (OPC Commands, OPC Complex Data, OPC Batch, and OPC Data eXchange) as well as a platform-independent but less performant specification for Web Service-based communication (OPC XML-DA) have been developed. Finally, OPC Security, and OPC Common Definitions are more general specifications focusing on defining the usage of security issues and basic aspects being important for the majority of the other specifications.

Several reasons led to the development of OPC UA. The ten main drivers have been summarized in [6]. Briefly, they are concerned with the discontinuation of COM/DCOM, DCOM limitations, OPC communication across firewalls, use of OPC on non-Windows platforms, high-performance OPC communication via Web Services, unified data model, support of complex data structures, process data communication without data loss, increased protection against unauthorized data access, and support of method calls.

[1] `www.opcfoundation.org`

The UA specification consists of 13 different parts of which some have already been adopted by the International Electrotechnical Commission as standard series IEC 62541. Briefly, they address a general overview [10], an extensive security concept [11], the Address Space model [12], abstract services [13], an information model [16], technology mappings [14], profiles [19], data access [15], data monitoring [20], method calls [17], historical data access [18], and discovery [21] as well as aggregation[2] functionalities for servers.

OPC UA realizes a server-client architecture following the Service-Oriented Architecture (SOA) paradigm. The layered communication architecture is based on two technology mappings for the communication (Web Services and a binary format) and modeling rules for data modeling purposes. The next layer addresses basic, generic services used for the communication between servers and clients. The communication is divided into different access types being oriented towards the Classic OPC specifications. The two top layers include a domain-specific information model used to represent the accessed data and appropriate vendor-specific specifications (see Figure 12.1).

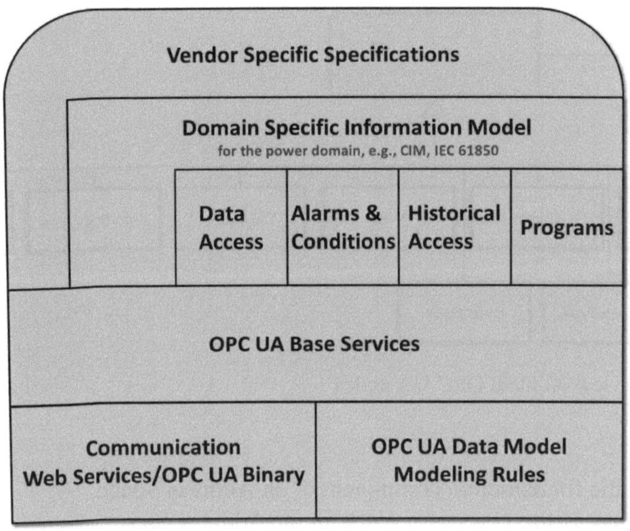

Fig. 12.1 OPC UA architectural overview

In the following sections, the two core aspects (information modeling and communication services) of the OPC UA are introduced. Moreover, key issues such as technology mappings, information security, and profiling are considered. Finally, the application of OPC UA in Smart Grids in accordance to appropriate data model mappings is analyzed.

[2] Part 13 is still under development.

12.2 Information Modeling

Besides the communication, modeling of information is the key aspect of OPC UA. The realized concept allows meta data annotation. Hence, information with known semantics can be exchanged instead of simple data. The abstract information model can be used in combination with both, standardized models and vendor-specific models. Former, however, provide a high level of interoperability because the data cannot only be exchanged in an interoperable way but also with clearly defined semantics. OPC UA provides a model that can be used in order to define specific information models. The model is called *Address Space*.

The Address Space mainly consists of a set of nodes being connected by certain references. The nodes a grouped in classes based on their meanings. Each class of nodes includes attributes describing them in detail. General attributes are for example *name* or a unique *NodeId*. Further attributes are related to the specific node classes. In parts 1 and 3 of the specification [10, 12], the main classes are depicted in Figure 12.2 and described as follows:

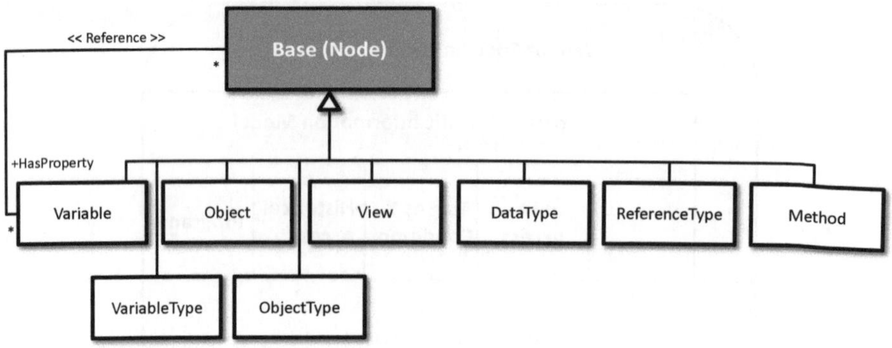

Fig. 12.2 Overview on main OPC UA nodes

- ***Nodes*** are the fundamental component of an Address Space.
- A ***NodeClass*** is the class of a *Node* in an Address Space. *NodeClasses* define the meta data for the components of the OPC UA object model. They also define constructs, such as *Views* that are used to organize the Address Space.
- An ***ObjectType*** is a *Node* that represents the type definition for an *Object*.
- ***Objects*** are *Nodes* that represent a physical or abstract element of a system. *Objects* are modeled using the OPC UA object model. Systems, subsystems, and devices are examples of *Objects*. An *Object* may be defined as an instance of an *ObjectType*.
- A ***ReferenceType*** is a *Node* that represents the type definition of a *Reference*. The *ReferenceType* specifies the semantics of a *Reference*. The name of a *ReferenceType* identifies how source *Nodes* are related to target *Nodes* and generally reflects an operation between the two, such as "A Contains B".

- A ***Reference*** is an explicit relationship (a named pointer) from one *Node* to another. The *Node* that contains the *Reference* is the source *Node*, and the referenced *Node* is the target *Node*. All *References* are defined by *ReferenceTypes*.
- ***VariableTypes*** are *Nodes* that represent the type definition for a *Variable*.
- A ***Variable:*** is a *Node* that contains a value. Two different types of *Variables* exist:
  - *Variables* that are the target *Node* for a *HasProperty Reference* are called ***Properties***. *Properties* describe the characteristics of a *Node*.
  - *Variables* that are representing the content of *Objects* are called ***DataVariable***.
- A ***DataType*** is represented by a *DataType Node*. The *DataType* is used together with the *ValueRank Attribute* to define the data type of a *Variable*.
- ***Views*** define a certain part of the Address Space in order to provide only the reasonable nodes for the user.
- A ***Method*** defines the signature of a method, which can be executed over OPC UA interfaces.

Complex *ObjectTypes* enable the definition of complex structures within the Address Space that can be reused for every application of the *ObjectType*. This concept is similar to the paradigm of object-orientation for programming languages. Accordingly, inheritance of *ObjectTypes* including adding attributes like *Variables* and *Methods* to the inherited objects is also supported. Based on the meta model, OPC UA defines a basic information model, which includes the *Base ObjectType*. The concepts introduced so far offer a number of possibilities in terms of creating extensions. On the one hand, simple extensions like defining sub-types of the *Base ObjectType* (including additional *Variables*, *Methods* or *Objects*) and sub-types of the *Base VariableType* (including sub-variables) can be made. On the other hand, more sophisticated extensions like specifying customized data types and reference types adding additional semantics to the nodes' relations, can be defined.

Concluding, in [9] the basic principles of information modeling with OPC UA are summarized as follows:

- Using object-oriented techniques including type hierarchies and inheritance
- Type information is exposed and can be accessed the same way as instances
- Full meshed network of nodes allowing information to be connected in various ways
- Extensibility regarding the type hierarchies as well as the types of references between nodes
- No limitation on how to model information by providing various extension mechanisms
- OPC UA information modeling is always done on the server-side; the model can be accessed and modified from OPC UA clients but an OPC UA client is not required to have an integrated OPC UA information model

Figure 12.3 contains an example applying information modeling using existing standardized information models based on OPC UA. The used notation is standardized

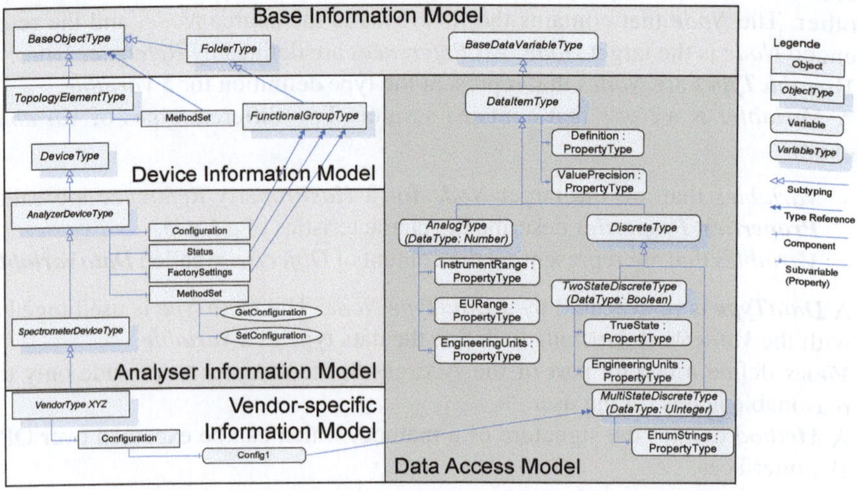

Fig. 12.3 Example of OPC UA information modeling [8]

by OPC UA and can be seen as the usage of Unified Modeling Language (UML) applying UML stereotypes. The base information model defines the base types, where the device information model (also called DI – Device Integration) provides common types to describe devices like flow meter or temperature sensor, but also controller or Intelligent Electronic Devices (IED). Derived from that, the analyzer device model (also called ADI – Analyzer Device Integration) defines types of analyzer devices having concrete characteristics with respects to the configuration or supported methods (e.g. GetConfiguration in Figure 12.3).

Each object of that type has the same structure and thus supports the same method. Finally, using subtyping vendor-specific extensions can be added, in Figure 1 indicated by VendorTypeXYZ. Variable types are already defined by the data access model which is part of the OPC UA specification. Those extended types offer the possibility to provide standardized information about the engineering unit (°C, bar, etc.) or the precision.

12.3 Communication Services

As already mentioned, UA communication is based on a server-client architecture. The Client-Application represents the code, which implements the functionalities of the clients themselves. The OPC UA Client Application Programming Interface (API) is used to invoke services (send requests and receive responses) from OPC UA servers. The API is an internal interface isolating the code of the Client Application from the UA communication stack.

Analogous to the clients, the servers also have an application and an API both having the same characteristics. However, the Server-Applications are more complex compared to the Client-Applications. Real objects represent physical (e.g., devices) or virtual (e.g., software) components. Furthermore, they include the Address Space with its nodes and views as described in Section 12.2. Monitored items are created by clients in order to monitor certain nodes of the Address Space or the real objects represented by the nodes, respectively.

The services that are used for the communication between servers and clients are specified in an abstract manner. In order to make them applicable, technology mappings have been developed (see Section 12.4). Part 4 of the UA specification [13] deals with abstract services for the communication. They were specified based on the "keep it short and simple" principle. As a result, a small set of easy-to-use services was developed. They were classified by their functionalities and aggregated to the following service sets:

- **Discovery Service Set** provides services being used for both, discovering endpoints, which are implemented by the servers and reading its security configurations.
- **SecureChannel Service Set** consists of those services that allow opening a communication channel to a server whereas the channel ensures for all exchanged messages their confidentiality, and integrity.
- **Session Service Set** addresses—within a session—the establishment of an application layer connection.
- **NodeManagement Service Set** comprises services for managing the nodes and references of Address Spaces. Thereby, they can either be newly added or deleted.
- **View Service Set** defines services used to navigate through an Address Space or a *View* as part of an Address Space.
- **Query Service Set** is intended to be used for issuing queries to servers. For the clients it is important to have knowledge of the servers' Address Space. Clients are allowed to access the data of the servers without knowledge of their logical schema.
- **Attribute Service Set** includes services that enable the access to the attributes of the nodes with Address Spaces.
- **Method Service Set** deals with the invocation of methods provided by servers.
- **MonitoredItem Service Set** provides services for creating, modifying, and deleting monitored items as well as for setting monitoring modes and triggering options.
- **Subscription Service Set** includes services, which cope with the subscription model. Similar to the monitored items, subscriptions can be created, modified, and deleted by the use of the services. Moreover, the publishing mode can be set and subscriptions can be transfered from one session to another.

The described services are the basis for the communication, which is usually started by the client opening a session. Therefore, certificates and authentication information have to be exchanged with the server. After establishing the session, the client can read and write data including current attribute values of nodes as well as other attributes providing meta data. If attributes support historicization of values that information can be accessed as well. The client can navigate through servers' Address Spaces by browsing or sending queries. Besides this relatively simple communication, the concept of subscription is also realized. This means that a client can subscribe for a certain value in order to get information about changes. As a result, data transmission is optimized because information is only transmitted if necessary. To further optimize the transmission, deadband and sampling rate approaches can be applied. Deadband is used to avoid notifications about negligible changes and the sampling rate specifies a time interval in which a change is maximally published. A third approach for communication is alarms and events. Events are transient and can be queried by subscriptions. They have an extensible set of arrays including, e.g., Message, Severity, and a unique identifier. Meta data concerning the event is accessible in the server. Alarms however, do have states, which can be read from the servers. For example, an event could be reaching a certain value and an alarm reaching a critical value.

12.4 Technology Mappings

In order to make the abstract UA services applicable while meeting specific requirements, different technology mappings for encodings and transport protocols have been specified. Figure 12.4 gives an overview on the defined mappings also taking into account security solutions.

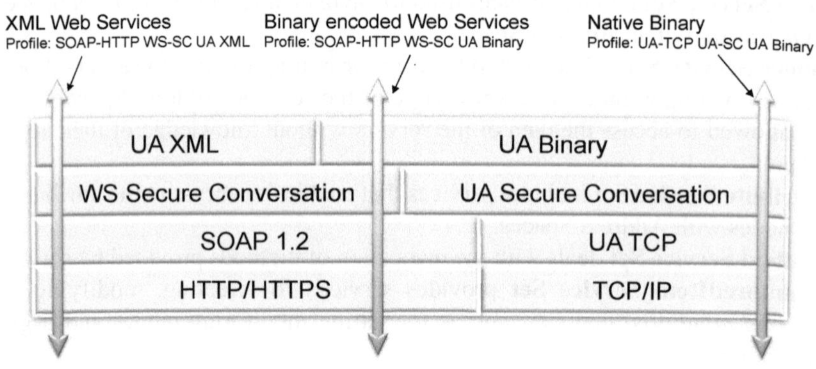

Fig. 12.4 OPC UA technology mappings [8]

The SOAP-mapping enables communication via Internet, e.g., for enterprise applications. The transferred data can either be encoded using Extensible Markup Language (XML) or an UA-binary format. Whereas, this mapping realizes a simple solution for XML-based communication, the second option focuses on both, fast and performant data transmission and encoding. An UA-TCP mapping based on the common TCP is specified. This option meets the communication requirements for control systems. In terms of security, the established *WS**-standards are used for the first option and an appropriately adapted version also for the second option. The OPC Foundation provides various development platforms (.Net 3.0, ANSI C, and JRE 5.0), which can be used to develop applications based on the abstract services and independent from the technology mappings. Afterwards, the existing mappings can be implemented or new ones are defined.

12.5 Profiles

The profile concept is addressed in part 7 of the OPC UA specification [19] and provides a means to realize servers with different complexities, which can be used for different use cases (e.g., embedded server for measuring one sensor or complex servers for SCADA-systems). Profiles are named groups of conformance units. Profile categories group profiles based on their major functionalities. Both types of OPC UA devices, servers and clients provide the names of the (multiple) profiles they support. Due to the fact that new profiles are expected to be defined in future, the list of profiles will continuously increase. The classification of profiles in profile groups is supposed to help users to understand the applicability of the profiles. For example, the profile category *Server* contains 94 conformance units, which specify a complete functional set for an OPC UA server. The profile *Discovery Client Facet* is a member of the profile group *Client* and consists of six conformance units. The included conformance units are either mandatory and or optional. Furthermore, each profile has an Uniform Resource Identifier (URI), which is part of a software certificate returned with the *CreateSession* service response.

Conformance units and conformance groups are part of the overall profile concept. A conformance unit is a specific set of features, which can be tested as a single entity. They are the building blocks of profiles. Conformance groups are groups of conformance units and their names are closely related to the introduced service sets. Conformance groups are used due to the large number of conformance units. They are only used for organizational structuring purposes. For example, the conformance group named *Discovery Services* comprises ten different conformance units that all deal with server endpoint discovery. Two of these conformance units are *Discovery Get Endpoints* and *Discovery Find Servers Self*, which both are members of the *Server* profile category.

12.6 Security

Security[3] plays a key role in terms of the practical use of a technology like OPC UA [23, 2]. For that reason and because OPC UA—in contrast to Classic OPC—is expected to be applied beyond the borders of closed automation applications, security was of major concern during the development of OPC UA. By the use of

[3] Smart Grid security in general, is further examined in Chapter 8.

UA, automation systems can now be connected to enterprise systems or the Internet. This leads to new challenges for security, for example malware like Stuxnet. Thus, in part 2 [11] of the UA specifications, six addressed protection goals are described as follows:

- **Authentication** for applications is realized by X.509 Certificates. For users token-based approaches (X.509v3 certificates, WS-SecurityToken, or username/password,) are intended to be used.
- In UA, **authorization** techniques are not explicitly specified so that authorization management of the product using UA has to be adapted.
- To meet **confidentiality** requirements, UA applies asymmetric encryption for key agreements and symmetric encryption for other messages exchanged between UA applications. Moreover, Cyber Security Management Systems (CSMS) and Public Key Infrastructures (PKI) are used for the encryptions.
- The objective of **integrity** is covered by asymmetric signatures for the key agreement process during the connection establishment and by symmetric signatures for other messages. The concept also relies on CSMS and PKI.
- **Auditability** is addressed by providing traceability of activities through log entries.
- The **availability** objective is closely related to the threat of message flooding, which is described below. If an attack tries to open more sessions than a server can handle, the server rejects sessions, which exceed its specified number.

Furthermore, the ten following threats to UA-systems have explicitly been identified [11]. Each of which affects one or more of the introduced protection goals:

- **Message Flooding** means either sending a large number of messages or sending one message, which includes a large number of requests. OPC UA faces those attacks minimizing the processing. (Impacts on: Availability)
- **Eavesdropping** aims at disclosing sensitive information and is met in OPC UA architectures by providing encryption mechanisms. (Impacts on: Confidentiality (directly) and also the other five objectives indirectly)
- **Message Spoofing** describes faking clients' or servers' messages on different layers of the protocol stack. UA, therefore signs messages, which also contain IDs for sessions, secure channels, and requests as well as the correct sequence number. (Impacts on: Integrity and Authorization)
- **Message Alteration** comprises capturing and modifying messages, which are afterwards being sent to servers or clients in order to gain unauthorized access to the attacked system. Again, signatures are applied to be checked for any changes and if necessary being rejected. (Impacts on: Integrity and Authorization)
- **Message Replay** means capturing messages and forwarding them later to servers or clients without changes. Again, UA uses several IDs like to counter message spoofing attacks and furthermore adds timestamps and sequence numbers to request and response messages. (Impacts on: Authorization)
- **Malformed Messages** are messages with an invalid structure or invalid data values being sent to servers or clients. UA simply checks whether messages have

the right form and values are within their ranges. (Impacts on: Integrity and Availability)

- **Server Profiling** means to get any information about a server to later use this information for further attacks. For example, messages can be sent to somehow draw conclusions from the responses. UA servers thus only provide very limited information to clients, which have not been identified before. (Impacts on: all six objectives)
- **Session Hijacking** requires knowledge about session IDs from current sessions. Manipulated messages can be injected in order to take over the session. Due to the security context like the secure channel applied to UA communications the context has to be compromised before a hijacking could take place. (Impacts on: all six objectives)
- **Rogue Servers** are either malicious servers or unauthorized instances of real servers. Hence, UA uses certificates for application instances and moreover PKI techniques. (Impacts on: all objectives except Integrity)
- **Compromising User Credentials** concerns information like user names, passwords, certificates or keys. Therefore, UA encrypts user credentials, which are sent over the network. (Impacts on: Authorization and Confidentiality)

Like the services and the information model, the security architecture is also basically generic. Hence, it is possible to select suitable implementations for the chosen technology mappings. Based on the mapping, the security goals are addressed on different levels. The UA security concept distinguishes between application, communication, and transport layer. On the application layer, sessions are established by the aforementioned session services, which in turn are based on secure channel services from the communication layer. Based on the technology mapping, well-established mechanisms are chosen to fulfill the protection goals. Secure Sockets Layer/Transport Layer Security (SSL/TLS) like specifications just as WS Security, WS Secure Conversation, XML Encryption, and XML Signature.

12.7 Power Domain-Specific Data Modeling

In order to make the OPC UA applicable for Smart Grid ICT-architectures, domain specific data models have to be integrated. More precisely, the highly recommended CIM (see Chapter 6) and IEC 61850-based data models (see Chapter 7) [26, 27] as well as the IEC 61131-3 for industrial control programming should be taken into consideration. The general applicability of the OPC UA for the power domain has been discussed—based on representative Smart Grid use cases—in [8] and [3].

12.7.1 IEC 61970/61968

The mapping of CIM to OPC UA is already discussed within the IEC who are working on a draft version of the mapping (IEC 61970-502-8 [5]). The electronic model of the CIM is developed using UML (Unified Modeling Language) and is published by the CIM Users Group[4] and the IEC. The data model includes several main packages with different functionalities. These packages include sub-packages and classes with attributes and associations. This set of abstract classes, attributes, and associations represents physical objects like cables and abstract objects like connectivity nodes. Altogether, in version 13 the model consists of 45 packages, ≈ 900 classes, ≈ 870 associations, and ≈ 2650 native attributes [28].

For the mapping, which is the basis for the implementation, the abstract CIM UML classes have been modeled as abstract UA *ObjectTypes*. The UA *Objects* represent specific instances of the abstract CIM classes. Concerning other different modeling decisions, basic design decisions have to be made. For example, in specific cases it has to be decided whether to model CIM attributes either as *Properties* or *Data Variables*. Furthermore, the CIM associations have to be modeled as *References*, but due to the cardinalities, a special UA *ReferenceType* has to be created. Up to this point the modeling is server independent. The next modeling steps for the server's architecture are specific. Especially the design of the *Views* is server specific and depends on the individual needs. A server can use the Views to provide different clients or groups of clients with access to parts of the model relevant to them. *Views* can also be used to deal with the concept of CIM profiles. The CIM is a very large data model and it is difficult and often not necessary to use the complete model for all purposes. To make the use of the CIM more applicable, one commonly uses profiles, which include only essential classes and associations of the CIM. In most cases, utilities extend the profiles with their own specific objects for special purposes [28].

The mapping is depicted in Figure 12.5, thereby the two branching arrows mean that the designer can choose between different options. This depends on the modeler's flavor of modeling and on the overall environment. Merging arrows only express that different CIM-elements may be mapped onto the same OPC UA structure.

CIM based Transformations—CIMbaT

The implementation of the introduced mapping concept between CIM elements and the OPC UA Address Space is supported by an Add-In for Sparx Enterprise Architect (EA)[5]. This Add-In is called CIM based Transformations (CIMbaT) [25]. The overall concept of CIMbaT follows a step-by-step wizard approach. The final result is an XML-file including all information required for generating a OPC UA server, compliant to the Software Development Kit (SDK) and *QuickStarts* provided by the OPC Foundation. Furthermore, a configuration file is used in order to maintain basic default settings like namespaces and prefixes for stereotypes. The UML stereotypes are utilized to add UA-specific information to the CIM model. The model is extended but not modified by the tool. Thus, all original CIM information is preserved.

---

[4] `http://cimug.ucaiug.org`

[5] `http://www.sparxsystems.com/`

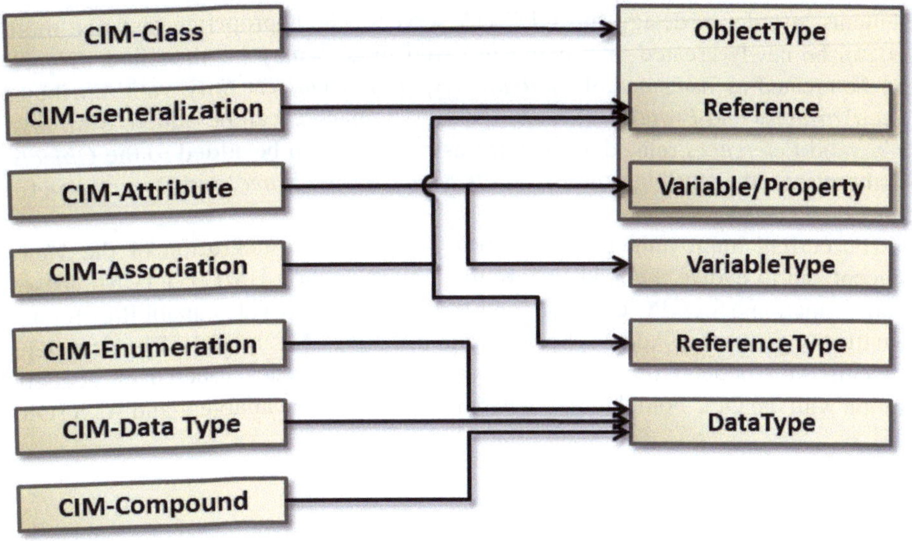

Fig. 12.5 Mapping of CIM elements onto the OPC UA Address Space [25]

The first step of the wizard deals with the default settings for the modeling process. After initially importing the model into internal data structures, design decisions regarding specific data type mappings and according to choices presented in Figure 12.5, can be made.

During the next step, design decisions for all CIM classes, attributes, associations, and data types can be made. The model is illustrated by a navigable tree-structure on the left-hand side. On the right-hand side, the settings for the appropriate element are provided. Here, the default settings previously made are preselected.

For CIM classes the attributes "IsAbstract" and "SupportsEvents" can be set *true* or *false*, respectively. In terms of CIM associations, the source and target role names are displayed. The direction of the association can be set, pointing from the source to the target class. It is possible to invert the direction. This design step is necessary due to the fact that CIM associations are undirected but in the Address Space the direction of a *Reference* has to be defined. Most settings are concerned with CIM attributes: referring to the descriptions of the default settings, the data types and the access rights can be selected from a list of valid values. Furthermore, it can be defined whether the attribute should be realized as *DataVariable* or *Property*. By setting the *Historizing* attribute to *true*, it is indicated that the server should actively collect data for the history of the attribute. Finally, it can be decided whether attributes should be mandatory or optional depending on the instantiation of the model. For all CIM elements, the appropriate descriptions are also accessible in order to support the model engineers with semantic information.

Beside the functionality dealing with the support of the defined mapping of the overall models, CIMbaT also allows to create instances of the Address Space elements in order to design full OPC UA servers. The appropriate instance models can be newly created or previously saved models may be modified. *Objects* can be created as instances of the *ObjectTypes*. For *Objects* only valid attributes (*DataVariables* and *Properties*) related to CIM semantics can be added. Similarly, only valid *References* related to the CIM associations can be added to the *Objects*. Furthermore, other attributes like *SymbolicName*, *BrowseName*, and *Description* for *Objects* can be set.

The current improvements of CIMbaT focus on the integration of the *View*-concept and to evolve the Add-In towards a more generic UMLbaT (UML-based Transformations) for OPC UA. The long-term goal is to provide capabilities for automatically generating Address Spaces from different UML models. The first steps will cope with models from the energy sector. The tool was developed in close cooperation with the OPC Foundation—where it will be made available—and ALSTOM [25].

12.7.2 IEC 61850

Unlike the CIM, the IEC 61850 does not only provide a simple data model but in addition mechanisms for the communication infrastructure like Functional Constraints (FC) for filtering the data, or timestamps and quality of the exchanged data. The IEC 61850 uses its own mechanisms to define its model and is not based on a pure object-oriented approach using UML (although the latest version of the IEC 61850 uses UML to document their approach [1]). Thus, the mapping cannot be performed in the same fashion as with the CIM.

Different approaches may be chosen to map the IEC 61850 model to an OPC UA information model [7]. For example, it has to be decided whether specific attributes of the IEC 61850 like quality and timestamp should be mapped the same way as all other attributes or handled specifically using the built-In OPC UA mechanisms having status codes and timestamps on each value. Furthermore, the FC defined for attributes in IEC 61850 could be made available in OPC UA using different modeling alternatives. Here, one possibility for the mapping is introduced. For the introduced mapping, the following decisions were made and depicted in Figure 12.6:

- *Logical Node* (LN) classes as defined in IEC 61850-7-x are generally mapped onto UA *ObjectTypes*.
- *LNodeTypes* are generally mapped onto UA *ObjectTypes* sub-typing the LN Class.
- LN are generally mapped onto UA *Objects* as instances of *LNodeTypes*.
- LN *Data* as the attributes of LN are mapped onto UA *Objects*.
- *Common Data Classes* (CDC) are also generally mapped onto UA *ObjectTypes*.
- CDC *DataAttributes* as the attributes of CDC are mapped onto UA *Variables*.
- CDC *DataAttribute Types* are the types of the CDC attributes and mainly mapped onto existing UA standard data types like *Integer*, *Float*, and *String*.
- FC are mapped onto UA *Objects*.

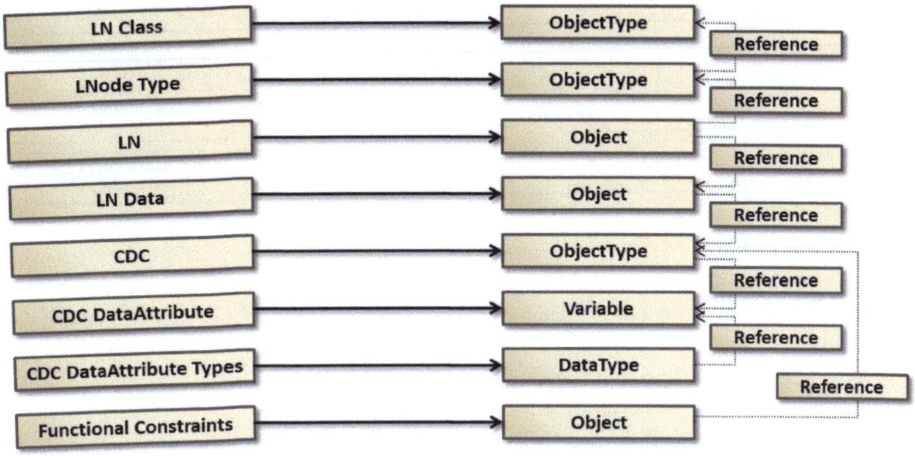

Fig. 12.6 Mapping of the IEC 61850 data structure onto the OPC UA Address Space [7]

In order to structure the objects three standard UA *ReferenceTypes* are used:

- *HasComponent* describes a part-of relationship between LN and its attributes as well as between CDC and its attributes. Furthermore, it is used for the grouping by FC.
- *Organizes* is used to group the CDC attributes by FC.
- *HasTypeDefinition* connects the LN attributes with the according CDC.

Mapping Example

The example [24] depicted in Figure 12.7 includes the LN "MMXU" and the CDC "MV" as well as their attributes. "MMXU" is a LN class, which shall be used for calculation of currents, voltages, powers, and impedances in a three-phase system. It is mainly used for operative applications. The CDC "MV" represents measured values. The focus is on only three attributes of the "MMXU": "TotVA" (Total Apparent Power), "TotVAr" (Total Reactive Power), and "TotW" (Total Active Power). Also for the "MV", a limited number of attributes, which can be divided by the FC is considered. FC shall indicate the services that are allowed to be operated on a specific attribute. The attributes "instMag" (magnitude of a the instantaneous value of a measured value), "mag" (current value of "instMag" considering deadband), "q" (quality of the measured value), "t" (timestamp of the measured value), and "range" (range of the current value of "instMag") belong to the FC "MX" (Measurands) and the attributes "subEna" (used to enable substitution), "subMag" (used to substitute the data attribute "instMag"), and "subID" (shows the address of the device that made the substitution) belong to the FC "SV" (Substitution). This is similar to modeling parameters for devices as defined in [10].

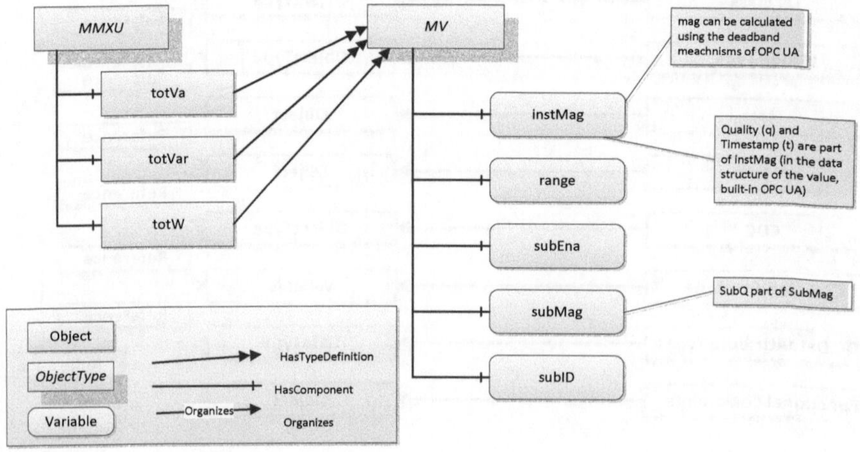

Fig. 12.7 Specific example for OPC UA and IEC 61850 mapping

The mapping shows that it is possible to expose the IEC 61850 model in OPC UA. By providing the LN class and the *LNodeTypes* in the UA Address Space, it is possible that pure OPC UA clients without any previous knowledge of the IEC 61850 can make use of the type model and design, e.g., specific HMI elements for any "MMXU".

12.7.3 IEC 61131-3

In a joint effort the OPC Foundation and PLCopen developed an OPC UA-based information model for IEC 61131-3 languages [4]. IEC 61131-3 standardizes programming languages for industrial automation and defines the common elements of the programming languages. The software model defines different resources with tasks and programs running in those tasks. Programs can be constructed out of function blocks. The standardized mapping of those concepts to an OPC UA information model is defined in [22]. The main purpose of the first version of the mapping is supporting the observation and operation of Programmable Logic Controller (PLC) programs. This includes reading and monitoring function block parameters and program variables as well as writing them. By using the type information rapid engineering is supported. For example, a user interface can be developed for a specific PLC program defined in IEC 61131-3. This user interface can be deployed to any PLC running this program without the need to reconfigure the user interface other than connecting to the representation of the program in the OPC UA server.

An example of the mapping is shown in Figure 12.8. The definition of the function block *CTU INT* realizing a counter is shown on the left-hand side. It is mapped to an OPC UA *ObjectType* inheriting from the generic *CtlFunction-BlockType* defined in [22] shown on the right-hand side. The variables of the function block are mapped to OPC UA *Variables* and the data types of the variables are mapped to the OPC UA *DataTypes* as defined in [22].

```
FUNCTION_BLOCK CTU_INT

VAR_INPUT
  CU: BOOL;
  R: BOOL;
  PV: INT;
END_VAR

VAR
  PVmax: INT := 32767;
END_VAR

VAR_OUTPUT
  Q: BOOL;
  CV: INT;
END_VAR

  IF R THEN
    CV := 0;
  ELSIF CU AND (CV < PVmax) THEN
    CV := CV + 1;
  END_IF ;
  Q := (CV >= PV);

END FUNCTION BLOCK
```

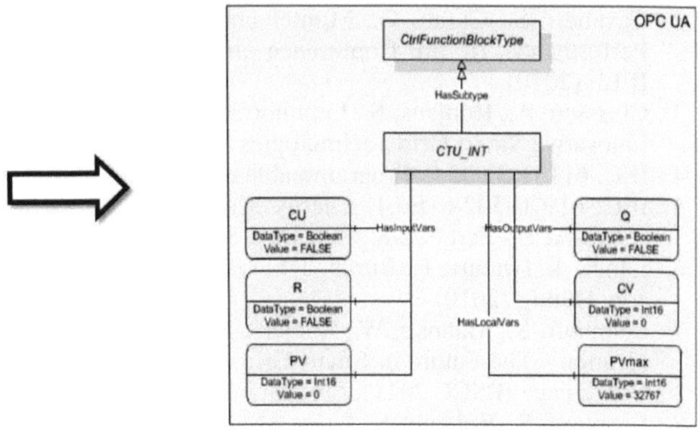

Fig. 12.8 OPC UA and IEC 61131-3 mapping [7]

12.8 Conclusion and Outlook

The OPC UA is developed by the OPC Foundation and partially standardized through IEC 62541. The UA is the successor of the established Classic OPC standards OPC DA, OPC A/E, and OPC HDA, which are mainly used for process automation by exchange of real-time plant data among control devices in industrial automation. New requirements like platform independence and Internet capability led to the development of the UA. The UA consists of 13 parts of which the parts three to six are in this context the most important ones. They specify an abstract data and information model (Address Space), which is the basis for a domain specific model, abstract service for server-client communications, and technology mappings, e.g., for a Web Service-based communication. The information modeling is done on server-side. Thus, clients consume information made accessible by servers. The abstract approach of the UA enables extensions of the application area, so that the focus is on general data exchange within any domain.

One of these new application areas is the future Smart Grid. In order to foster the realization of OPC UA-based communication, mappings have been introduced, which cover the most essential semantics for Smart Grid use cases. Besides the highly recommended CIM and IEC 61850, an integration with IEC 61131-1 has been introduced as well. Concluding, the OPC UA provides with all capabilities required to become a vital part of future ICT-architectures in the power domain.

References

1. Apostolov, A.: IEC 61850 Substation Configuration Language and Its Impact on the Engineering of Distribution Substation Systems. In: Congreso Internacional de Distribución Eléctrica (CIDEL 2010), pp. 1–6 (2010)
2. Cavalieri, S., Cutuli, G., Monteleone, S.: Evaluating Impact of Security on OPC UA Performance. In: 3rd Conference on Human System Interactions (HSI), pp. 687–694. IEEE (2010)
3. Claassen, A., Rohjans, S., Lehnhoff, S.: Application of OPC UA for the Smart Grid. In: Innovative Smart Grid Technologies (ISGT) Europe 2011 (2011)
4. IEC: 61131-3 ed2.0: Programmable controllers - Part 3: Programming languages (2003)
5. IEC: 61970-502-8 Ed.1: Energy Management System Application Program Interface (EMS-API) - Part 502-8: CIM Data Services (Draft) (2008)
6. Lange, J., Iwanitz, F., Burke, T.J.: OPC: From Data Access to Unified Architecture, 4th edn. Hüthig (2010)
7. Lehnhoff, S., Mahnke, W., Rohjans, S., Uslar, M.: IEC 61850 based OPC UA Communication - The Future of Smart Grid Automation. In: 17th Power Systems Computation Conference (PSCC 2011), Stockholm (2011)
8. Lehnhoff, S., Rohjans, S., Uslar, M., Mahnke, W.: OPC Unified Architecture: A Service-Oriented Architecture for Smart Grids. In: ICSE 2012 International Workshop on Software Engineering Challenges for the Smart Grid. IEEE (2012)
9. Mahnke, W., Leitner, S.H., Damm, M.: OPC Unified Architecture. Springer (2009)
10. OPC Foundation: OPC Unified Architecture Specification Part 1: Overview and Concepts - Release 1.01 (2009)
11. OPC Foundation: OPC Unified Architecture Specification Part 2: Security Model - Release 1.01 (2009)
12. OPC Foundation: OPC Unified Architecture Specification Part 3: Address Space Model - Release 1.01 (2009)
13. OPC Foundation: OPC Unified Architecture Specification Part 4: Services - Release 1.01 (2009)
14. OPC Foundation: OPC Unified Architecture Specification Part 6: Mappings - Release 1.00 (2009)
15. OPC Foundation: OPC Unified Architecture Specification Part 8: Data Access - Release 1.01 (2009)
16. OPC Foundation: OPC Unified Architecure Specification Part 5: Information Model - Release 1.01 (2009)
17. OPC Foundation: OPC Unified Architecture Specification Part 10: Programs - Version 1.01 Draft (2010)
18. OPC Foundation: OPC Unified Architecture Specification Part 11: Historical Access - Release Candidate Version 1.01 (2010)
19. OPC Foundation: OPC Unified Architecture Specification Part 7: Profiles - Version 1.01 Draft 1 (2010)
20. OPC Foundation: OPC Unified Architecture Specification Part 9: Alarms & Conditions - Release 1.00 (2010)
21. OPC Foundation: OPC Unified Architecture Draft Specification Part 12: Discovery - Version 1.02 (2011)
22. OPC Foundation, PLCopen: OPC UA Information Model for IEC 61131-3 - Release 1.00 (2010)
23. Renjie, H., Feng, L., Dongbo, P.: Research on OPC UA Security. In: 5th IEEE Conference on Industrial Electronics and Applications (ICIEA), pp. 1439–1444. IEEE (2010)
24. Rohjans, S., Piech, K., Mahnke, W.: Standardized Smart Grid Semantics using OPC UA for Communication. IBIS - Interoperability in Business Information Systems 6(10) (2011)

25. Rohjans, S., Piech, K., Uslar, M., Cabadi, J.F.: CIMbaT - Automated Generation of CIM-based OPC UA-Address Spaces. In: IEEE SmartGridComm 2011, Brussels (2011)
26. Rohjans, S., Uslar, M., Bleiker, R., González, J.M., Specht, M., Suding, T., Weidelt, T.: Survey of Smart Grid Standardization Studies and Recommendations. In: First IEEE International Conference on Smart Grid Communications (2010)
27. Uslar, M., Rohjans, S., Bleiker, R., González, J.M., Suding, T., Specht, M., Weidelt, T.: Survey of Smart Grid Standardization Studies and Recommendations - Part 2. In: IEEE Innovative Smart Grid Technologies Europe (2010)
28. Uslar, M., Rohjans, S., Specht, M., González, J.M.: What is the CIM lacking? In: Innovative Smart Grid Technologies Conference Europe ISGT Europe 2010 IEEE PES (2010)

Chapter 13
Market Communication

José M. González and Michael Specht

Abstract. Market communication is a core activity within the energy sector as it integrates the business view (cash flow) with the technical view (power and gas flow) of the energy sector. To ensure an efficient operation of the energy supply chain from generation to its use at the customer side, standards regarding market operation are essential. Within this chapter the need for standards regarding market communication is motivated and an overview on current market communication standards is provided. Finally, a short overview and summary on the introduced standards is presented.

13.1 Market Communication and the Need for IT Standards

Market communication in this context refers to electronic business transactions between business partners and market roles (like network operators, suppliers and traders) in the energy industry (power and gas likewise) to support the supply of energy from its generation to the provision of consumers with energy. One market participant can take more than one role but might be limited to certain roles due to national regulation.[1] Market communication can be seen as a network of relationships between market participants comprised of different business processes. These business processes aim at consumers and producers of energy and related services, whereas the consumer can be an end-user, a producer, or even both.

The motivation for market communication is clearly depicted in the following citation from the European Federation of Energy Traders [23]:

José M. González · Michael Specht
OFFIS – Institute for Information Technology, Escherweg 2, 26121 Oldenburg, Germany
e-mail: {jose.gonzalez,michael.specht}@offis.de

[1] For example German regulation forces companies in the energy industry to unbundle activities related to transmission and generation and hence force them to establish independent companies.

M. Uslar et al.: *Standardization in Smart Grids*, POWSYS, pp. 211–228.
DOI: 10.1007/978-3-642-34916-4_13

"Communication is an essential key to the successful integration of business processes. Successful communication requires that the communicating parties speak the same language. This fact is as important in electronic communication as it is in face to face communication."[23]

To achieve the use of one common language, IT communication standards should be applied. The use of standards in the market communication aims at reducing the costs of application integration in internal and intra-company business processes. Standardizing the data exchange can, apart from cost reductions, lead to a significant increase of efficiency.

In case of regulatory changes in the energy market, or if extension or replacement of technical components are necessary, standards can minimize the effort, secure the ability to upgrade, or find suitable components and hence minimize investment risks.

Most standards leave room for interpretation when coming to their application, so testing on standard conformity plays a particular role. This topic is explained in detail in Chapter 9.

One important point at using standards is that standards are not absolutely stable and without mistakes, but are constantly improved in new releases. Part of these improvements can be the integration of new applications and technologies, innovations on devices, new market participants or even completely new market roles into the standards. Additionally, it is necessary to bear in mind that different market participants have varying speeds in migrating new versions of a standard, so it must be possible to communicate with different versions of the same standard at a time.

Altogether when considering IT standards for market communication the following topics have to be considered, see Figure 13.1:

- **Business partners and market roles:** market communication involves the exchange of data between companies through the whole energy supply chain. This requires the consideration of different market roles and business partners.
- **Unique identifiers:** as different companies are taking part in market communication processes, correct and unique identification of abstract or artificial (like documents) and physical (like technical equipment) objects is needed.
- **Data formats:** to enable an efficient data exchange standardized data formats, that are used by all participants, are recommended.
- **Processes:** to ensure a transparent communication between the participants in the energy market processes should be specified.
- **Test cases:** as standards often provide room for interpretation and implementations may differ, test cases should provide support to check for standard compliance.
- **Lifecycle management:** as standards constantly evolve support for different versions of standards through lifecycle management is required.

The remainder of this chapter is organized as follows. After motivating the need of standards for market communication in this section, Section 13.2 will introduce several standards and related standards developing organizations.

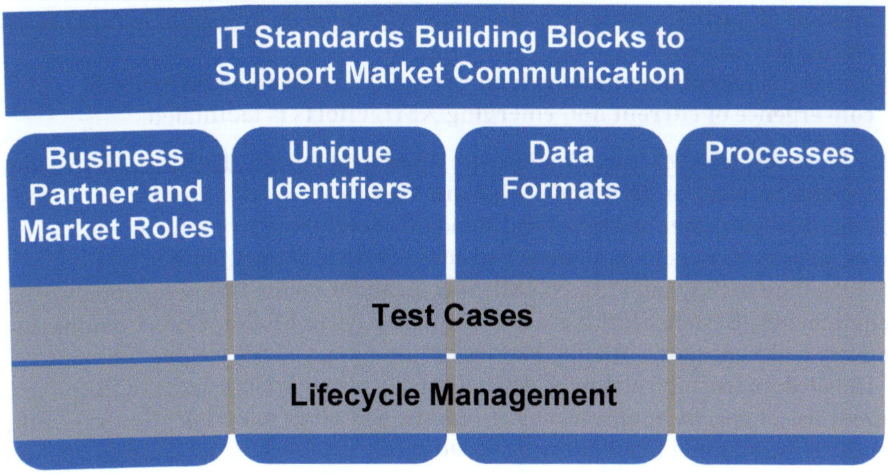

Fig. 13.1 Core building blocks of IT standards to support market communication

13.2 IT Standards and Standards Developing Organizations for Market Communication

In this section, first of all, basic methodologies and data formats are introduced, see Sections 13.2.1 and 13.2.2. Based on this the major standardization developing organizations like OASIS or IEC and associations like EFET, ENTSO-E or ebIX as well as their core specifications and models are presented. Finally in Section 13.3 an overview regarding IT standards and their corresponding responsible organizations is provided.

13.2.1 Electronic Business Using eXtensible Markup Language

In 1999, the Organization for the Advancement of Structured Information Standards (OASIS) and the United Nations/ECE agency CEFACT started an initiative to provide data specifications including XML standards for business processes, core data components, collaboration protocol agreements, messaging, registries, and repositories [32].[2] This was the basis for a suite of specifications to conduct business over the internet called Electronic Business using eXtensible Markup Language (ebXML).

According to [25], the ebXML specifications provide a framework which tries to preserve the investments in Business Processes using Electronic Data Interchange (EDI) by providing an architecture which uses the new technical capabilities of the Extensible Markup Language (XML).

ebXML provides the following key characteristics [32]:

- a **globally developed open XML-based standard** built on electronic business experiences,

[2] See `http://www.ebxml.org/`

- **all kind of parties** irrespective of size are supported to engage in internet-based electronic business,
- parties can **complement and extend current EC/EDI investment**, and
- **convergence of current and emerging XML efforts** is facilitated.

To achieve this, ebXML specifications are developed for the ebXML infrastructure by experts building on EDI knowledge and experiences [32]. Further on, OASIS collaborates with other initiatives and standards development organizations and engages industry leaders to participate and adopt ebXML infrastructure.

According to [33] and [25], ebXML specifications on the following topics exist: Requirements, Business Process and Information Meta Model, Core Components, Registry and Repository, Trading Partner Information, and Messaging Services.

Detailed information about the several specifications are provided in [33]. For a current list of specifications, technical reports, and other material[3]. Several of the standards introduced in the following sections rely on ebXML, see Section 13.3 for an overview.

13.2.2 UN/CEFACT Modeling Methodology

The UN/CEFACT Modeling Methodology (UMM) is according to [37] a UML modeling approach aiming at supporting collaboration between business partners. It addresses a service-oriented architecture (SOA) based development of low cost software to enable small and medium sized companies as well as emerging economies to engage in e-business. Therefore, it focuses on developing a global choreography of inter-organizational business processes and their information exchanges providing technology independent models using the UML syntax. Based on the platform independent UMM models, services that need to be realized in a service-oriented architecture can be identified.

UMM provides a formal description technique to specify a class of business transactions having the same business goal based on the Business Operations View (BOV) of the ISO/IEC 14662 "Open-EDI reference model". The ISO/IEC 14662 BOV only considers aspects of business transactions dealing with business decisions and commitments among organizations, implementation specific technological aspects of Open-EDI are not taken into account.

All UMM models are based on the UMM meta model outlined in Figure 13.2. The UMM meta model contains a set of modules which divide the meta model into functional levels.

In the following the different partition levels are described [37]:

- ***Base*** includes the fundamental elements, that are shared overall.
- ***Foundation*** specifies the core concepts of the UMM. Here, the minimal methodology to develop a UMM compliant business collaboration model is defined.

---

[3] `http://www.ebxml.org/specs/index.htm`

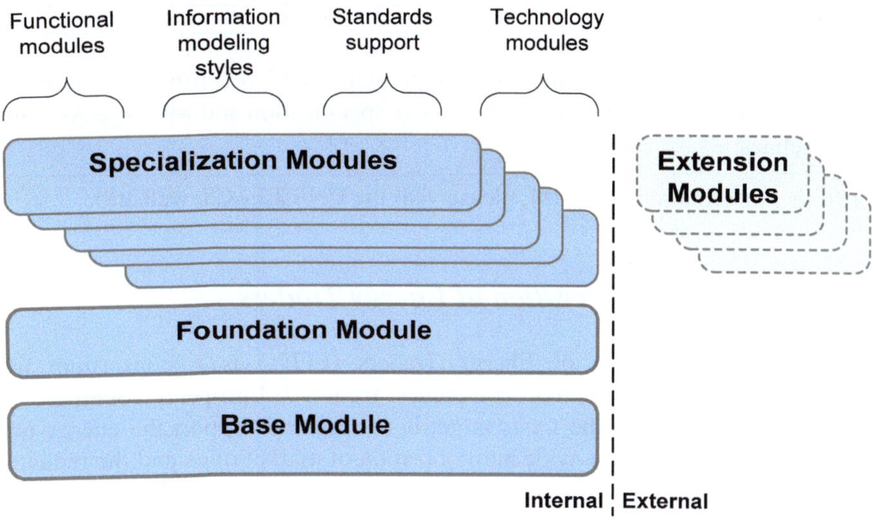

Fig. 13.2 UMM meta model module structure [37]

- ***Specialization*** defines modules with add-on concepts to address specialized types of analysis. Specialization modules might be later included into the foundation module.
- ***Extension*** focuses on the same purpose as specialization modules, but are developed and maintained by organizations outside of UN/CEFACT.

Each partition is based on the underlying one and enhances it. Therefore, concepts defined at the bottom can be used by higher portions, but not vice versa.

As collaboration is the core focus of UMM, UMM introduces the concept of a UMM business collaboration model (see foundation module). According to UMM, a UMM business collaboration model contains three main views [37]:

- ***Business Requirements View (bRequirementsV)*** comprises existing knowledge from stakeholders and business domain experts. Here use cases are used to describe intra- and inter-organizational business processes on a rather high level. As a result use cases are developed in the language of the business experts and stakeholders. The bRequirementsV typically contains a map/categorization (packages) of business processes (use cases).
- ***Business Choreography View (bChoreographyV)*** includes the overall choreography between collaborating business partners in an inter-organizational business process. The bChoreographyV itself contains further views like the Business Transaction View—where business actions of each partner when sending and receiving business information are specified—and the Business Collaboration View—here all requirements of business collaboration use cases and their participating authorized roles are documented. The requirements of both views serve as basis for their implementation in a SOA collaboration architecture.

- ***Business Information View (bInformationV)*** contains business information artifacts which allow the definition of information exchange in a document-centric approach. Regarding the modeling approach UMM recommends to use the UN/CEFACT Core Components Technical Specification and Message Assembly Guidelines.

For further information on UMM, please visit the UN/CEFACT website[4].

13.2.3 European Federation of Energy Traders

The European Federation of Energy Traders (EFET) is a consortium with more than 100 energy trading companies from 27 European countries, see `http://efet.org/`. The focus target is to ease and support the energy trading with the help of Europe-wide harmonization of market rules and the reduction of market entry barriers [22]. The standards developed in the EFET framework, especially the basic agreements, are designed by energy traders for over-the-counter (OTC) trading with gas and electrical energy between European wholesalers [26].

The steadily increasing trade volume, count of transactions per day, and the rising count of trading partners, are demanding a nearly fully automatized electronical market communication to ensure an efficient market with low transaction costs. This cannot be obtained with existing communication devices.

Because there are no standardized processes in the market communication domain, EFET focuses on the standardization and harmonization of energy trading contracts, business use cases, and electronical data exchange. EFET strives for a seamless IT support (straight-through processing) at the electronical data exchange, which needs a specification of message structures and message content and the message exchange itself for energy trading processes.

Instead of covering all possible energy trading processes, EFET describes the "Electronic Confirmation and Matching" process and creates extensions for additional processes. The use of peer-to-peer communication is a main assumption within the EFET process description. Each communication partner has to manage the message exchange independently. The alternative method of using an agency based network, which holds a third party to manage these, has been abandoned.

The following enumeration shows the main EFET Standards:

- **EFET Standard Documentation General Agreements (GA):** the so called General Agreements comprises standardized energy trading contracts for buying and selling electricity [18], Gas [20], emission allowances [21], and coal. Only trades on the wholesale market with physical content are described. Financial products are not part of the specification. The energy trading contracts are the basic part for the processes. The EFET Framework is used on a widespread basis in Europe, especially in the energy domain [26].

[4] `http://www.unece.org/cefact/umm/umm_index.html`

- **Electronic Confirmation and Matching (eCM) [24]:** the Electronic Confirmation and Matching specification describes a standardized business process to confirm trading transactions between trading partners. Particularly the message structure, content, and how the messages have to be exchanged is part of the description. Each trading partner has to check the contracts independently. The purpose of standardizing is to ensure the reporting duty and to ease the maintenance of the trading system. The eCM documents use the ebXML Message Service Specification v2.0[5] as transport protocol. The description of the process is done by UML diagrams (primarily use case diagrams and sequence diagrams), whereas the data structure is described with XML.
- **Electronic Position Matching (ePM) [19]:** this part handles the matching of trading positions between different trading partners based on the previous mentioned eCM standard. The ePM aims at gaining an overview of the trading positions (accounts receivable and payable) towards the trading partner. This overview can be essential to estimate the risks.
- **Electronic Settlement Matching (eSM) [23]:** parallel to the Electronic Position Matching the Electronic Settlement Matching extends the Electronic Confirmation and Matching standard with a feature to compare invoices between different trading partners.

Regarding the application of EFET standards several requirements have to be considered. EFET requires to carry out tests and get a certificate for the IT-System before allowing to participate at the market. These tests should ensure the interoperability between different systems from different vendors. To conduct such a test, a testing iteration with correct data and an additional testing iteration with intentional wrong data will be done. The wrong data test is conducted to increase the sturdiness at rejecting false messages. At the moment two vendors are offering eCM and the including tests, whereas ePM is only offered by one vendor and no tests are necessary. If another vendor offers ePM functionality in the future, the interoperability test features have to be added on to the existing system.

To increase the operational suitability, additional validation tests shall be provided. These tests could ensure the consistence of the data and increase the data quality. Such elements have to be used early to minimize the effort for adjustments.

13.2.4 European Network of Transmission System Operators for Electricity

The European Network of Transmission System Operators for Electricity (ENTSO-E) organization is a federation of European Transmission System Operators, which was created to generate a common electricity market. ENTSO-E is the successor of ETSO (European Transmission System Operators). The following goals are in the focus of ENTSO-E:

[5] See `http://www.ebxml.org/specs/ebMS2.pdf`

- simplify the European electricity market,
- research and development of general guidelines for the harmonization and creation of rules to improve the grid operation and quality of service, and
- create solutions for scientific and regulatory questions, which are interesting for transmission system operators(TSO).

To reach the afore mentioned goals, different task forces have been created. One of the task forces is charged with the work on concepts for the electronical data exchange between transmission system operators, market participants, and distribution network operators. The following artifacts have been developed:

- **ENTSO-E Energy Identification Coding Scheme (EIC) [15]:** identification of market participants (like network operator, trader, or supplier), regional structures (like accounting grid and control area), as well as meter points in the energy market. The management and release of new codes is decentralized under the organization of the local issuing offices.
- **ENTSO-E Modeling Methodology (EMM) [9]:** the Method was developed to support an automated data exchange in business processes, based on a harmonized role model[6]. Among others a rule set to derive XML messages out of UML models was derived using the ETSO Core Components. Here the UN/CEFACT Modeling Methodology(UMM) as well as parts of ebXML were applied.
- **ENTSO-E Electricity Market Harmonized Role Model [17]:** a role model developed by EFET and ebIX to identify roles and sections and to harmonize terms in the domain.
- **ENTSO-E Core Components (ECC) [16]:** descriptions of elements in the energy domain and a unique naming.
- **ENTSO-E Scheduling System (ESS) [10]:** specification of the exchange of schedules (like day ahead or intra day schedules), created to support the development of schedule managing systems. Currently this specification is used in several countries in Europe like Germany, Spain, Austria, Italy, and others.
- **ENTSO-E Settlement Process (ESP) [12]:** this specification helps to develop an IT application for market players that can exchange electricity market settlement information, such as finalized schedules and imbalance reports within a given balance area.
- **ENTSO-E Reserve Resource Process (ERRP) [11]:** specification for the information exchange for reserve resource tendering, planning and activation in the balance management process.
- **ENTSO-E Capacity Allocation and Nomination (ECAN) [13]:** this specification defines an information exchange of the transmission capacity rights allocations and nominations within scheduling processes.
- **ENTSO-E Market Data Exchange Standard (MADES) [14]:** MADES defines a decentralized common communication platform based on international IT protocol standards. On the one hand software interfaces to exchange electronic data are described. On the other hand basic services for, e.g., authentication, encryption, message tracking, and message logging are specified.

[6] See `https://www.entsoe.eu/resources/edi-library/`

Further cooperations between the IEC TC 57 for improving the electronic data exchange are already ongoing. The previous described specifications are steadily under development. One of the main topics is to develop a CIM profile known as CIM Market Extension (CME), based on the IEC 61970 CIM (further described in Chapter 6).

13.2.5 Energy Business Information eXchange

The European forum for energy Business Information eXchange (ebIX) aims at standardizing the electronic data exchange within the European downstream energy sector [8]. ebIX was founded in 2003 as European standardization body out of members of the Ediel Nordic Forum[7]. At the moment, companies and organizations from Austria, Belgium, Denmark, Germany, the Netherlands, Norway, Slovenia, Sweden, and Switzerland are ebIX members [8]. ebIX focuses on the exchange of administrative data for the European market for gas and power addressing the retail and wholesale (downstream) market.

Within this ebIX follows the rules and regulations of the European Union and is continuously in contact with other standardization bodies like IEC[8], EFET[9], ENTSO-E[10] and EURELECTRIC[11]. In contrast to EFET and ENTSO-E, which address specific market roles, ebIX has a more general approach. In Figure 13.3 an overview of the ebIX organization, its project groups and links to other organizations is illustrated.

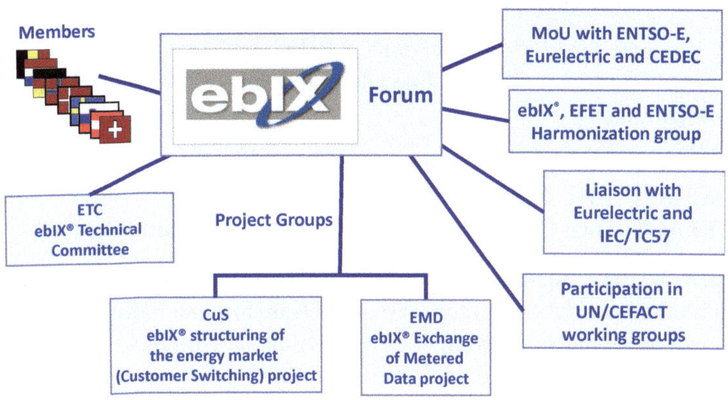

Fig. 13.3 The ebIX organization based on [8]

7 See http://www.ediel.org

8 See http://www.iec.ch/

9 See Section 13.2.3.

10 See Section 13.2.4

11 See http://www.eurelectric.org/

The core goals of ebIX are according to [8]:

- the standardization of the electronic data exchange regarding administrative data in the European energy market,
- the coverage of the whole supply chain (from wholesale to retail) for power and gas considering the requirements of multi-utility companies;
- and finally to become the de-facto standardization body for data exchange in the European downstream energy market.

ebiX aims at achieving these goals through

- the consideration of the requirements of the different market participants and markets taking local requirements into account,
- the development of practical solutions which are applied voluntary,
- and the application of guidelines and specifications of the European Union.

In the following the basic documents of ebIX with regard to electronic data exchange are introduced [8]:

- ***The Energy Business Domain Model [6]*** comprises a functional description of the European energy market with links to market roles. This domain model serves as basis for further models and descriptions.
- ***The Harmonized Role Model [17]*** is developed and maintained collaboratively by ENTSO-E, EFET, and ebIX. Here an aligned UML market role model of actors in the energy domain (like traders, balancing responsible parties, and customers) is provided as pdf document and UML project file. The current version is available on the ebIX website[12].

For ebIX process descriptions the ebIX Modelling Methodology [7] is applied which is based on proven methods like UML and the following UN/CEFACT technical specifications:

- **UN/CEFACT Modeling Methodology (UMM)**[13]
- **UN/CEFACT XML Naming and Design Rules (NDR)**[14]
- **UN/CEFACT Core Components Technical Specification (CCTS)**[15]

Within ebIX the ebIX Technical Committee (ETC) is responsible for the introduced modeling methods and the maintenance of ebIX documents. This working group is also responsible for harmonization of ebIX documents with other organizations and supports the use of ebIX standards. Regarding the application of the ebIX framework the ebIX vendor group is of major importance, as this group focuses on the transfer of standardization work into practice. As data exchange involves several parties and different bodies are involved in e-buisness standards, harmonization is needed to leverage existing work. Figure 13.4 outlines the different liaisons of ebIX with other standards development organizations and associations.

---

[12] `http://www.ebix.org`

[13] See `http://www.untmg.org/`

[14] See `http://www.unece.org/cefact/`

[15] See `http://www.untmg.org/`

Fig. 13.4 ebIX links to other standardization bodies [8]

Based on the previously introduced modeling method and models the following projects were initiated which describe selected processes in the energy market:[16]

- ***Customer Switching Project CuS*** describes processes regarding the exchange of structured data focusing on the automated exchange of business documents. This mainly comprises processes like customer switching (e.g. change of energy supplier or balancing supplier) and maintenance of master data. CuS aims at defining common standards for data interchange to enable the automation of processes.
- ***Exchange of Metered Data EMD*** comprises the exchange of metered data in the European electricity market. EMD developed a model for the upstream European energy market which relies on the overall ebIX deregulated model and serves as common basis for the different parties involved. At the moment, the EMD model contains the following individual models:
  - Measure for imbalance settlement
  - Measure for reconciliation
  - Measure for billing
  - Measure determine meter read for switch
  - Settle reconciliation
  - Collected data
  - Labeling

[16] Further descriptions and reports regarding ebIX projects can be found at `http://www.ebix.org`

Apart from the above, process descriptions Guidelines for Digital Signatures (DigSig) for encryption and digital signature within the European energy sector are provided, see `http://www.ebix.org`.

In addition to the above presented technology independent models and documents, also descriptions and tooling for a message based data exchange using specific data formats like Electronic Data Interchange for Administration, Commerce, and Transport (EDIFACT) and XML are provided.

Currently, the ebIX framework is implemented in Austria, Belgium, Denmark, Finland, Germany, the Netherlands, Norway, Sweden, and Switzerland [8]. Here millions of messages are exchanged between hundreds of actors. The main implemented processes using the ebIX framework are [8]:

- structuring related data exchange—including customer switching—,
- exchange of metered data—for settlement, reconciliation and billing—,
- bidding on the Nordic power exchange (NordPoolSpot), and
- scheduling (in some countries).

13.2.6 IEC 62325 Standard

Working Group 16 (Deregulated energy market communications) of the Technical Committee 57 of the International Electrotechnical Commission (IEC) describes the IEC 62325 standard [30] as a framework for market communications using ebXML as base technology. The standard describes communications between e-business applications in the deregulated energy market, but with focus on the communication links between network operators and other market participants like traders and power plant operators. In Figure 13.5 the energy supply chain as understood by the IEC 62325 from generation to consumption is illustrated.

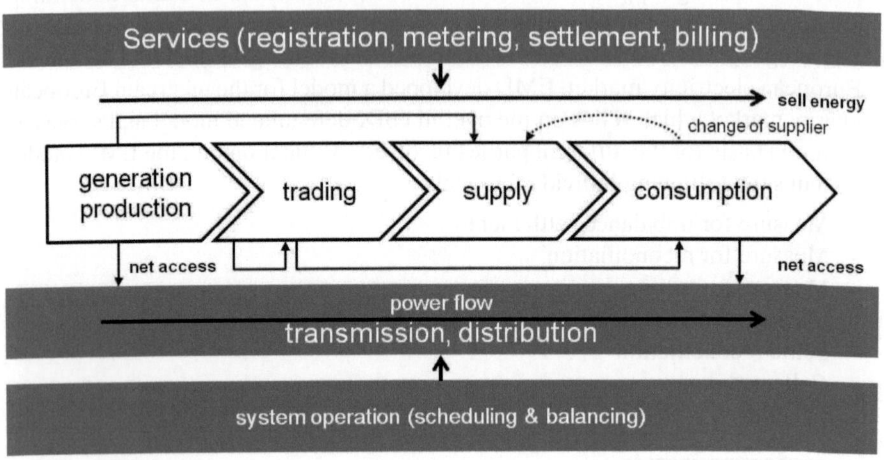

Fig. 13.5 Energy supply chain based on [30]

The Interface Reference Model (IRM) specified in IEC 61968, which describes the communication between distribution network management and external applications, is especially taken into account.[17] However, the IEC 62325 itself specifies not a new standard as such, but the ebXML standard as well as the UN/CEFACT specification along with other referenced standards are used in the context of energy markets. The goal is to present an alternative solution to existing communications based on EDIFACT, X12, and proprietary solutions.

In the following the core parts of the IEC 62325 are listed.

- **TR Part 101: General guidelines and requirements [30]:** this part describes an example of business models for the electricity models on the basis of open die reference models according to ISO/IEC 14662.
- **TR Part 102: Energy market model example [27]:** the UMM (UN/CEFACT Modelling Methodology), which is used in TR 101.
- **TR Part 501: General Guidelines and ebXML [28]:** this part of IEC 62325 provides general guidelines on how to use the ebXML technology and architecture in energy markets including migration scenarios.
- **TS Part 502: Profile of ebXML [29]:** this part of IEC 62325 specifies an energy market specific messaging profile based on the ISO 15000 series. The profile is intended to provide the basis for system configuration.

The currently published standards are limited in the description of using the ebXML technology. There are more parts in development which are using additional technologies, which should be published in between 2012 and 2014. The goal should be to provide an open and technology independent framework.

13.2.7 OASIS Smart Grid Suite of Standards

The Organization for the Advancement of Structured Information Standards (OASIS) is a not-for-profit consortium, that aims at the development, convergence and adoption of open standards for the global information society [36].

Therefore, OASIS produces worldwide standards for several domains like security, cloud computing, SOA, the Smart Grid, emergency management, and others. Over 5,000 participants representing over 600 organizations and individual members in 100 countries are engaged within OASIS. For further information on OASIS see `http://www.oasis-open.org`.

Currently, the OASIS Smart Grid Suite of Standards consists of three standards which are described according to [34] in the following. These standards address three of the National Institute of Science and Technology (NIST) Priority Action Plans (PAPs) 3, 4, and 9[18]. Hereby requirements across and within domains of the Smart Grid are considered.

---

[17] See Section 6.6 in Chapter 6.

[18] For further information on NIST PAPs see `http://collaborate.nist.gov/twiki-sggrid/bin/view/SmartGrid/PriorityActionPlans#Priority_Action_Plans_PAPs_Descr`

- ***Energy Market Information Exchange (eMIX)*** provides a standardized methodology to describe energy products that might be traded in a competitive marketplace and includes an information model for energy and market information [35]. It addresses NIST PAP0 3: "Develop Common Specification for Price and Product Definition", which claims the need for a common specification for defining products with characteristic attributes and price information. Through eMIX specialized technical vocabulary energy products and corresponding price information can be specified in such a way that buyers or sellers can easily form offers using attributes that all parties understand [35].
- ***WS-Calendar*** provides a common information model and vocabulary for calendaring and scheduling. It supports NIST PAP04: Develop Common Schedule Communication Mechanism for Energy Transactions by defining a common vocabulary.
- ***Energy Interoperation*** specifies interactions for conveying price quotes (like clearing prices) and tenders (offers to buy or sell) supporting EMIX schedules with energy product information. Several information objects in Energy Interoperation are based on EMIX and WS-Calendar definitions. It addresses NIST PAP 09: Standard DR and DER Signals by specifying corresponding processes and common vocabularies to use.

For an overview on the dependencies between the three OASIS Smart Grid standards see Figure 13.6 and [34]. EMIX makes use of WS-Calendar's information model for calendaring and scheduling and in turn is used by Energy Interoperation services for communicating price and product information. Energy Interoperation uses the information models provided by WS-Calendar and EMIX to implement demand-response, distributed energy resource interactions, and transactive energy interactions.

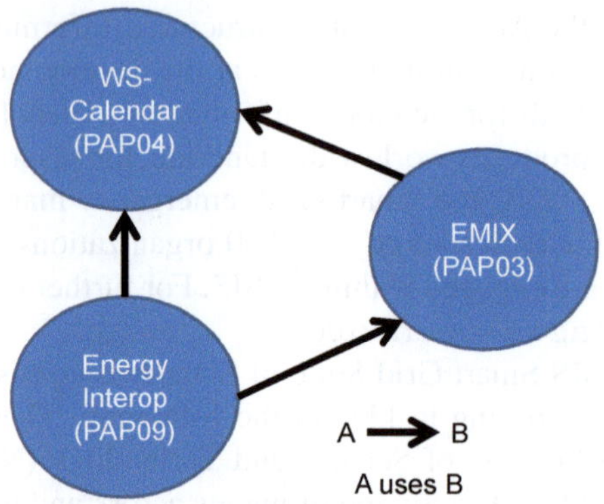

Fig. 13.6 Dependency graph for EMIX, Energy Interoperation, and WS-Calendar based on [35]

13.2.8 *German Market Communication Specifications*

As market communication is subject to national regulation the German guidelines and specifications are outlined exemplarily in the following.

In Germany the Federal Network Agency for Electricity, Gas, Telecommunications, Post and Railway (BNetzA), see `http://www.bundesnetzagentur.de`, is the institution in charge for defining regulatory guidelines and specifications for market communication. Due to liberalization and to increase competition, in 1998 the German energy market energy supply was broken down into generation, wholesale trading, transport, distribution, and energy supply [31]. In consequence regulation of the transport and distribution networks as natural monopoly was decided.

At the moment, different network access ordinances for electricity (StromNZV [4]) and gas (GasNZV [5]) are available. In addition, two different departments within the BNetzA are responsible for regulatory issues for electricity (BK7) and gas (BK6). Therefore, regulation for electricity and gas is similar, but several differences still exist. Currently two core guidelines regarding electricity and gas exist, which determine the regulatory authority for supplier switching processes, **"GPKE"** [1] and **"GeLi Gas"** [2]. They specify uniform business processes for legal supplier relationships. In addition the data formats to be used and the information exchange is defined, too. For an overview on the most important laws and ordinances see [3].

GPKE includes the following processes and assigns corresponding message types [1]: Supplier Switching, End of Supply, Begin of Supply, Supply in Case of Loss of Supplier, Meter Data Transfer, Change of Master Data, and Request for Business Data.

GeLi Gas focuses on similar processes for gas. For both, electricity and gas, uniform business processes and data formats were developed. Based on this the following message types for electronic data exchange are specified:

- ***UTILMD:*** data exchange during supplier switching (Change of Master Data)
- ***MSCONS:*** transfer of customer energy consumption (Meter Data)
- ***INVOIC:*** billing messages between network operator and supplier
- ***REMADV:*** advice of settlement
- ***REQDOC:*** document requests
- ***CONTRL:*** confirmation of receipt and syntax control
- ***APERAK:*** confirmation of acceptance

13.3 Summary

In this chapter an overview on the previous introduced organizations and related standards for market communication was presented, see Figure 13.7.

Altogether, even though several organizations exist, a trend towards using of common international standards is observable. Currently international standards are gaining momentum and therefore the different organizations try more and more to leverage existing standardization efforts mainly from the IEC and adapting them to national or European requirements instead of developing new standards.

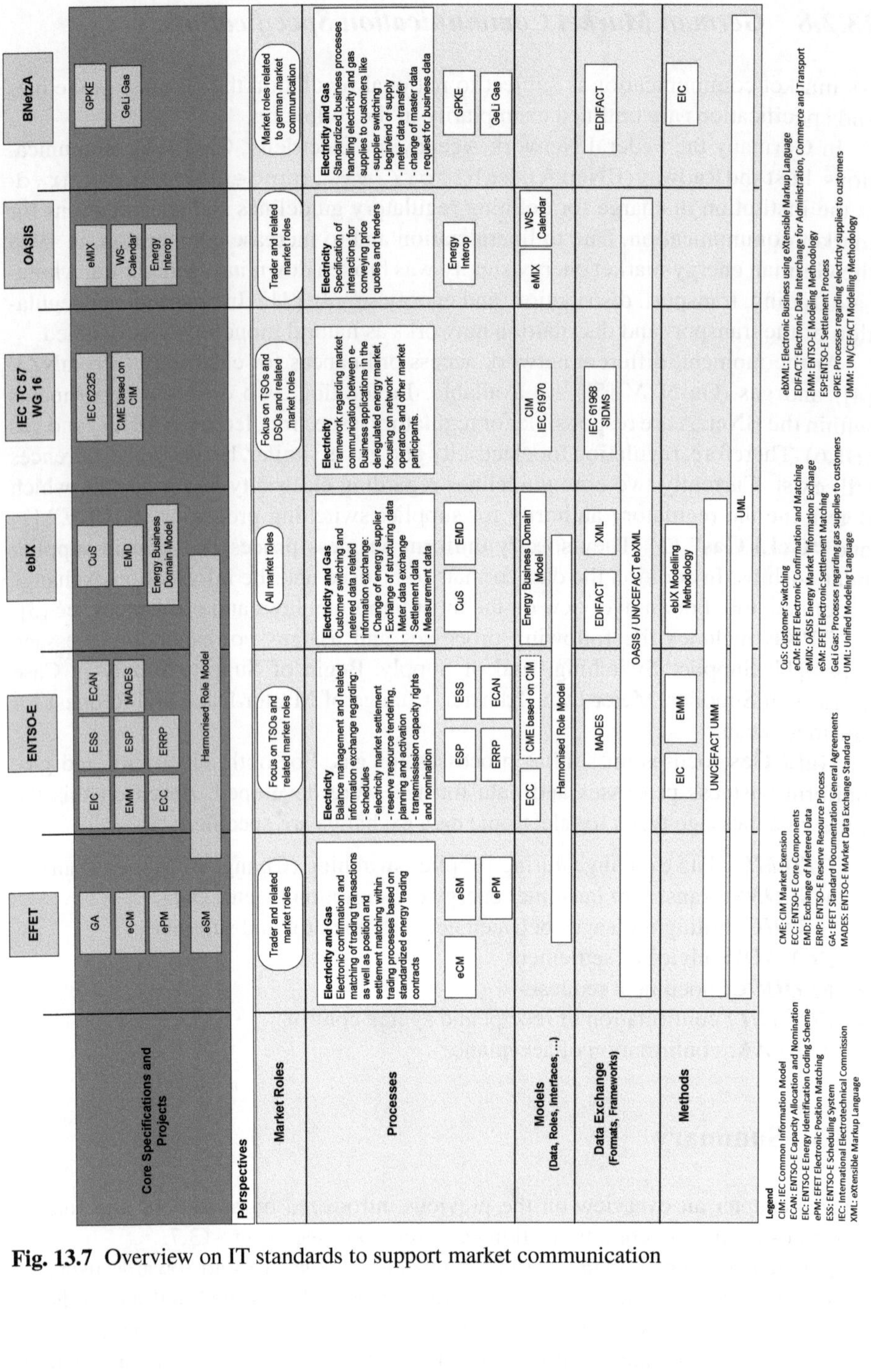

Fig. 13.7 Overview on IT standards to support market communication

References

1. Bundesnetzagentur: Anlage zum Beschluss BK6-06-009: Darstellung der Geschäftsprozesse zur Anbahnung und Abwicklung der Netznutzung bei der Belieferung von Kunden mit Elektrizität (Geschäftsprozesse zur Kundenbelieferung mit Elektrizität, GPKE). Tech. rep. (2006)
2. Bundesnetzagentur: Anlage zu dem Beschluss BK7-06-067: Geschäftsprozesse Lieferantenwechsel Gas (GeLi Gas). Tech. Rep. (August 2007)
3. Bundesverband der Energie- und Wasserwirtschaft (BDEW): Survey of the most important laws, ordinances, specifications, guidelines and recommended action on the subject of electricity business processes Status: 18 December 2008 Document Essential contents in regulatory terms. Tech. Rep (December 2008)
4. Deutscher Bundestag: Verordnung über den Zugang zu Elektrizitätsversorgungsnetzen (Stromnetzzugangsverordnung - StromNZV), pp. 1–14 (2008)
5. Deutscher Bundestag: Verordnung über den Zugang zu Gasversorgungsnetzen (Gasnetzzugangsverordnung - GasNZV) (2008)
6. Energy Business Information eXchange (ebIX): European energy market domain model. Tech. rep. (2004)
7. Energy Business Information eXchange (ebIX): ebIX Rules for the use of UN/CEFACT Modeling Methodology (UMM version 2). Tech. rep. (2012)
8. Energy Business Information eXchange (ebIX): Overview of ebIX work Goals and results to date (2012)
9. ENTSO-E: ENTSO-E Modelling Methodology for the Automation of Data Interchange of Business Processes (EMM) (2003)
10. ENTSO-E: ENTSOE Scheduling System (ESS) Implementation Guide Version: 3 Release: 3, pp. 1–113 (April 2009)
11. ENTSO-E: ENTSO-E Reserve Resource Process (ERRP) (2010)
12. ENTSO-E: ENTSO-E Settlement Process (ESP) Implementation Guide, pp. 1–48 (2010)
13. ENTSO-E: ENTSO-E Capacity Allocation and Nomination System (ECAN) Implementation Guide, pp. 1–217 (2011)
14. ENTSO-E: MADES Communication Standard. Tech. rep. (2011)
15. ENTSO-E: The Energy Identification Coding Scheme (EIC) Reference Manual (2011)
16. ENTSO-E: Entso-E General Codelist For Data Interchange (2012)
17. ENTSO-E, European Federation of Energy Traders (EFET), energy Business Information eXchange (ebIX): The Harmonised Elecricity Market Role Model. Tech. rep. (2011)
18. European Federation of Energy Traders (EFET): EFET - General Agreement Concerning the Delivery and Acceptance of Electricity Version 2.1(a) (2007)
19. European Federation of Energy Traders (EFET): EFET Electronic Position Matching (ePM) Version 1.1. Tech. rep. (2007)
20. European Federation of Energy Traders (EFET): EFET General Agreement Gas: Version 2.0 (a). Tech. rep. (2007)
21. European Federation of Energy Traders (EFET): Allowances Appendix (POWER) (2008)
22. European Federation of Energy Traders (EFET): The Past and Future of European Energy Trading. Tech. rep. (2008)
23. European Federation of Energy Traders (EFET): EFET Electronic Settlement Matching (eSM): Version 0.1: Final Draft. Tech. rep. (2009)
24. European Federation of Energy Traders (EFET): EFET Electronic Confirmation Matching. Tech. rep. (2011)
25. Grangard, A., Eisenberg, B., Nickull, D., Boseman, A., Barret, C., Brooks, D., Casanave, C., Technologies, D., Cunningham, R., Traffic, M., Command, M., Ferris, C., Microsystems, S., Kacandes, P., Ketels, K.: ebXML Technical Architecture Specification v1.0.4. Tech. rep. (2001)

26. Horstmann, K.P., Cieslarczyk, M.: Horstmann-Cieslarczyk: Energiehandel: Ein Praxishandbuch, Heymanns, Köln (2006)
27. IEC: 62325-102 DTR Ed.1: Framework for energy market communications - Part 102: Energy market example model (2004)
28. IEC: 62325-501 DTR Ed.1: Framework for energy market communications - Part 501: General guidelines of using ebXML (2004)
29. IEC: 62325-502 DTS Ed.1: Framework for energy market communications - Part 502: Profile of ebXML (2004)
30. IEC: IEC/TR 62325-101 ed1.0: Framework for energy market communications - Part 101: General guidelines (2005)
31. Kurth, M.: The Changing Structure of the Utility Industry from the Perspective of Regulation Authorities. In: Bausch, A., Schwenker, B. (eds.) Handbook Utility Management, pp. 193–206. Springer, Berlin (2009)
32. OASIS: ebXML - Enabling A Global Electronic Market
33. OASIS: ebXML Documentation Roadmap v0.93. Tech. rep. (2001)
34. OASIS: EMIX 1.0 and the OASIS Smart Grid Suite of Standards. Tech. Rep. (October 2011)
35. OASIS: EMIX Overview Version 1.0. Tech. Rep. (October 2011)
36. OASIS: OASIS (2012)
37. UN/CEFACT: UML Profile for UN/CEFACT's Modeling Methodology (UMM) - Foundation Module - Version 2.0 Technical Specification. Tech. rep. (2011)

Chapter 14
Looking Ahead: The Future of Smart Grid Communications and Standardization

Mathias Uslar

Abstract. As this book has already outlined, standardization for Smart Grids has become an important aspect and gained much attention in the community. The need of a joint effort by Information and Communication Technology, automation, and utilities can only be achieved using standards. Different aspects, as introduced, exist and cover large and a vast number of topics of the Smart Grid infrastructure. Within this section, we outline the general status of standardization, existing flaws and current trends.

14.1 The Good

As this book, probably even the first chapter should have shown, the topic of standardization has grown from a niche to a more important emerging topic. When systems have to be integrated to fulfill both the political, economical and ecological vision, seamless integration between the systems from the different vendors with the existing infrastructure has to be enforced. Interoperability has become one of the biggest issues discussed when systems from different domains have to be integrated. While traditionally, this topic has been very much in focus for the ICT domain (including EAI (Enterprise Application Integration), EMB (Enterprise Message Bus), SOA (Service-Oriented Architecture), and Cloud paradigms), the idea was quite new to the automation and utility domain.

Since 2008, a lot of work has been done in the context of Smart Grid standardization. Starting with the NIST Framework and Roadmap for Smart Grid Interoperability Standards, Release 1.0 [6], [4] and BMWi e-Energy [3] studies, later on with the IEC roadmaps [7], first papers actually addressing relevant standards for the Smart Grid use cases were created. Fortunately, the results of those studies which

Mathias Uslar
OFFIS – Institute for Information Technology, Escherweg 2, 26121 Oldenburg, Germany
e-mail: uslar@offis.de

M. Uslar et al.: *Standardization in Smart Grids*, POWSYS, pp. 229–233.
DOI: 10.1007/978-3-642-34916-4_14

were created without coordination or discussion between them, led to similar results identifying the core standards for the future Smart Grid. With those core standards on the very table for discussion from a national viewpoint, different national roadmaps have adopted the view for their national core standards. This in fact, was a big achievement as the nucleus for both Smart Grid communication and modeling was identified.

Soon after, the idea that standards are not a solution themselves without being applied as best-practice became apparent. The NIST 2.0 framework as well as the European initiatives from the M/490 Smart Grid mandate [5] focus on properly using, extending and adopting the core standards. The objective of the mandate is to develop or update a set of consistent standards within a common European perspective as well as integrating a variety of digital computing and communication technologies and electrical architectures, associated processes, and services, that will achieve interoperability and will enable or facilitate the implementation in Europe of the different high level Smart Grid services and functionalities as defined by the Smart Grid Task Force, which should be flexible enough to accommodate future developments. Building, Industry, Appliances, and Home Automation are out of the scope of this mandate; however, their interfaces with the Smart Grid and related services have to be treated under this mandate.

This initiative was a huge leap forward in Smart Grid standardization as, first time after the Joint-Working-Group Smart Grids report, communication (ETSI), electrotechnicians (IEC) and automation (ISO) worked alongside a storyline covering the integration of their best-practices using ICT. In addition, a link to the NIST initiatives was built with the Smart Grid Advisory Committee (SGAC) groups and regular discussions about architectures, roles, actors, domains, and use cases [1].

One particular example of a successful exchange is the European adoption of the IEC/PAS 62559 for the CEN/CENELC/ETSI M/490 Sustainable Processes (SP) Group. This provides easier exchange of Smart Grid use cases, their functional and non-functional requirements and makes standardization much easier. However, those initiatives are slowly picking up pace. With methodologies and vocabularies aligned, collaboration is much easier. But there is still no such thing as Smart Grid heaven (except so some failed solutions) which will be, a bit more provokingly, shown in the next sections.

14.2 The Bad

Unfortunately, the things have only improved in certain aspects but the situation is still far from being perfect. Different problems still exist when having to cope with Smart Grid standardization. After having read this short introductory book, you probably know more about it than before. But still, each individual chapter has only been a very short overview and introduction into thousands of pages from different standards and specifications. If you have to fully understand each of the individual standards, it will probably take you a lot of time to get to know the connections and technical implementations to make them work together seamlessly. So, achieving overall interoperability can only be achieved when all the needed stakeholders can work together, align on a common vocabulary and method engineering, and take their time to sort out the needed glitches between the standards to work together seamlessly. However, there is no real initiative besides the First Set of Standards Group from the EU mandate M/490 and the NIST Priority Action Plans to actually address gaps in standards, introduce new work item proposals (or parts) to change the existing standards and have work leaders cover them [4].

Unfortunately, this work is driven mainly from the national committees and their agenda. A lot of the PAPs from the ANSI are not really useful for the European mandate or may even be conflicting with European solutions. In addition, there is no European view on the Smart Grid as the use cases and regulatory issues still differ very much—one good example is how data privacy or smart metering are implemented and rolled out. In addition, a lot of "Me, too" initiatives have been started from companies trying to push their products and solutions into both Smart Grid and standardization. If important players from ICT try to enter the utility market, their influence is rather high and the overall amount of stakeholders is increasing while the knowledge about Smart Grids pretty much stays the same. Important aspects like the integration of CIM and IEC 61850, which is a rather crucial aspect when trying the integrate solutions from SCADA and field automation perspective, are neglected.

A lot of time in standardization is wasted discussing about input from third parties who try to influence the Smart Grid agenda with their products without actually solving the known problems which have been well documented. Another aspect is that there are too little experts actually both available in terms of the sheer number and their spare time. With the need to cover travel expenses, other parties than vendors (communication and automation) and research are less involved in the standardization process. Too little utilities actually want to cover the expenses for sending their employees to standardization. On the other hand, their requirements and experiences are really needed to come up with meaningful solutions and experience related to other than research projects and product development. Those problems are still easy to be solved in comparison with the ones described in the next section.

14.3 The Ugly

With the good improvements since 2008, and the open gaps and issues discussed, there are still some harder problems to be solved [2]. Standardization is not done as an end in itself - or at least it should not be. If there is no clear use case, open problem or new innovation, it is hard to actually come up with a new meaningful standard. One web comic from XKCD[1] clearly depicts another problem, the re-invention of the wheel or the not-invented here syndrome.

[1] Licensed under CC, http://creativecommons.org/licenses/by-nc/2.5/

Fig. 14.1 On standardization ©XKCD.com under CC 2.5

The core standards for the Smart Grid actually focus very much on the basic inter- and intra-utility infrastructure. At the borders, new standards emerge. Smart Metering is one of the aspects different projects and regulators focus on to make end customers become aware of their energy consumption, consume less energy or be more efficient. On the other hand, this causes actually new costs for the end customer who has to pay for the new infrastructure (at least up to a certain degree). In addition, this domain is pretty much dominated by different standards and regulatory regimes in different countries—so there is no one-size-fits-all solution. The purely technical problem of metering is extended by the different requirements from the legal side. Data privacy is considered to be a serious issue and has strong impacts on the acceptance and cost model of the Smart Grid applications. In addition, the utilities have to face that customers will be more than a load, a participant they might have to deal with even though more beneficial Smart Grid use cases exist.

A lot of emphasis in Smart Grid standardization is put on this very aspect while the return-on-investment, except for energy savings is still pretty unclear. In addition, the problem has worsened because different standards and proprietary solutions dominating certain markets exist, but also there are different groups having to deal with the interface between the Smart Meter (Smart Home, Smart Gateway, Multi-Utility Gateway and Home Gateway can be used synonymously in this aspect). IEC TC 57 WG 21, IEC PC 118 and initiatives like OGEMA or EEBUS strive for the lead on this interface and, of course, also have to deal with the legal aspects. Those interfaces are dominated from the ICT and communications side and should be more supported by utilities. Coping with this, and, in addition electric vehicles, is a huge issue which has a strong impact on Smart Grid standardization, but can only be solved the regulatory or political level as those solutions are mainly influenced by their ROI.

An additional and most striking aspect of standards is to taylor them for a specific use case. This is called profiling a standard [8]. For the existing standards, most of the time only blue-prints or building blocks exist—there is a need for profiles for the individual data models, communication mappings, and functions of the individual standards to test against. Otherwise, the standards would be pretty useless.

14.4 Recommendations and Trends

Having discussed the current approaches, the next steps for Smart Grid standardization can be outlined or at least put into some recommendations we see from our perspective as authors of this book.

- Concentrate to consolidate the participating stakeholders in Smart Grid standardization, mainly get more utilities involved and provide meaningful use cases
- If you are a manager responsible for product development or utility operations, consider to spent some budget and time for your experts to participate in standardization. It pays off because you save on trainings and get hands-on knowledge in return. However, this also need a knowledge-sharing organizational culture.
- Keep track on the European initiatives and the gaps to be closed. In addition, you will soon find out about a meaningful set of standards and a well-documented architecture and methodology to deal with.
- Focus on adopting certain profiles of a standard for your use cases, try to improve them and provide feedback back to standardization.
- Concentrate on your strategy to adopt core standards for your business first and later try to enlarge if needed. The need can come from internal IT costs which are lowered by standards like CIM, from regulation or from best practices of customers of OEMs.
- Try to make a harmonized IT strategy for both commercial and process IT and focus on a meaningful architecture management
- Try to make yourself familiar with latest trends such as OPC-UA to properly integrate legacy systems and future Smart Grid standards

References

1. CEN, CENELEC, ETSI: JWG Report on Standards for the Smart Grid. Tech. rep. (2010)
2. DIN: Die deutsche Normungsstrategie aktuell (2009)
3. DKE: Die deutsche Normungsroadmap E-Energy/Smart Grid (2010)
4. EPRI: Report to NIST on the Smart Grid Interoperability Standards Roadmap. Tech. rep., EPRI (2009)
5. European Commission: M/490 Standardization Mandate to European Standardisation Organisations (ESOs) to support European Smart Grid deployment (2011)
6. NIST: NIST Framework and Roadmap for Smart Grid Interoperability Standards (2010)
7. SMB Smart Grid Strategic Group (SG3): IEC Smart Grid Standardization Roadmap (2010)
8. Uslar, M., Rohjans, S., Bleiker, R., González, J.M., Suding, T., Specht, M., Weidelt, T.: Survey of Smart Grid Standardization Studies and Recommendations - Part 2. In: IEEE Innovative Smart Grid Technologies Europe (2010)

Appendix A
CIM Package Description

Table A.1 IEC 61968-11 data model packages and content

IEC 61970-301	
Common	This package contains the information classes that support distribution management in general.
Assets	This package contains the core information classes that support asset management applications that deal with the physical and lifecycle aspects of various network resources.
AssetInfo	This package is an extension of Assets package and contains the core information classes that support asset management and different network and work planning applications with specialized AssetInfo subclasses.
Work	This package contains the core information classes that support work management and network extension planning applications.
Customer	This package contains the core information classes that support customer billing applications.
Metering	This package contains the core information classes that support end device applications with specialized classes for metering and premise are network devices, and remote reading functions.
LoadControl	This package is an extension of the Metering package and contains the information classes that support specialized applications such as demand-side management using load control equipment.
PaymentMetering	This package is an extension of the Metering package and contains the information classes that support specialized applications such as prepayment metering.

Table A.2 IEC 62325 data model packages and content

IEC 61970-301	
MarketCommon	This package contains the common objects shared by both MarketManagement and MarketOperations packages.
MarketManagement	This package contains all core CIM Market Extensions required for market management systems.
MarketOperations	This package contains all core CIM Market Extensions required for market operations systems.

Table A.3 IEC 61970-301 data model packages and content

IEC 61970-301	
Domain	The domain package define primitive datatypes that are used by classes in other packages. Stereotypes are used to describe the datatypes.
Core	Contains the core PowerSystemResource and ConductingEquipment entities shared by all applications plus common collections of those entities. Not all applications require all the Core entities.
DiagramLayout	This package describe diagram layout. With layout it is meant how objects are arranged in a coordinate system rather than rendered.
OperationalLimits	The OperationalLimits package models a specification of limits associated with equipment and other operational entities.
Topology	An extension to the Core Package that in association with the Terminal class models Connectivity, that is the physical definition of how equipment is connected together.
Wires	An extension to the Core and Topology package that models information on the electrical characteristics of Transmission and Distribution networks,
Generation	This package contains packages that have information for Unit Commitment and Economic Dispatch of Hydro and Thermal Generating Units, Load Forecasting, Automatic Generation Control, and Unit Modeling for Dynamic Training Simulator.
LoadModel	This package is responsible for modeling the energy consumers and the system load as curves and associated curve data. Special circumstances that may affect the load, such as seasons and daytypes, are also included here.
Outage	An extension to the Core and Wires packages that models information on the current and planned network configuration. These entities are optional within typical network applications.
AuxiliaryEquipment	Contains equipment which is not normal conducting equipment such as sensors, fault locators, and surge protectors.
Protection	An extension to the Core and Wires packages that models information for protection equipment such as relays.
Equivalents	The equivalents package models equivalent networks.
Meas	Contains entities that describe dynamic measurement data exchanged between applications.
SCADA	Contains entities to model information used by Supervisory Control and Data Acquisition (SCADA) applications.
ControlArea	The ControlArea package models area specifications which can be used for a variety of purposes.
Contingency	Contingencies to be studied.
Statevariables	State variables for analysis solutions such as powerflow.

Appendix B
CIM RDF Topology

This section shows the complete RDF topology example from chapter 6.4.2.

Listing B.1 CIM RDF Topology

```
<?xml version="1.0" encoding="iso-8859-1"?>
<rdf:RDF xmlns:cim="http://iec.ch/TC57/2008/CIM-schema-cim13#"
xmlns:rdf="http://www.w3.org/1999/02/22-rdf-syntax-ns#">
  <cim:EnergySource rdf:ID="A_G">
    <cim:IdentifiedObject.name>Power Generator
       </cim:IdentifiedObject.name>
    <cim:activePower>400</cim:activePower>
  </cim:EnergySource>
  <cim:Terminal rdf:ID="A_Te1">
    <cim:Terminal.ConductingEquipment rdf:resource="#A_G" />
    <cim:Terminal.ConnectivityNode rdf:resource="#A_CN1" />
  </cim:Terminal>
  <cim:ConnectivityNode rdf:ID="A_CN1"></cim:ConnectivityNode>
  <cim:Terminal rdf:ID="A_Te2">
    <cim:Terminal.ConductingEquipment rdf:resource="#A_B1" />
    <cim:Terminal.ConnectivityNode rdf:resource="#A_CN1" />
  </cim:Terminal>
  <cim:BusbarSection rdf:ID="A_B1">
    <cim:IdentifiedObject.name>Busbar 1
       </cim:IdentifiedObject.name>
  </cim:BusbarSection>
  <cim:Terminal rdf:ID="A_Te3">
    <cim:Terminal.ConductingEquipment rdf:resource="#A_A1" />
    <cim:Terminal.ConnectivityNode rdf:resource="#A_CN1" />
  </cim:Terminal>
  <cim:ACLineSegment rdf:ID="A_A1">
    <cim:Conductor.length>2500</cim:Conductor.length>
    <cim:Conductor.r>0.3125</cim:Conductor.r>
    <cim:Conductor.x>0.28</cim:Conductor.x>
    <cim:Conductor.bch>235.45</cim:Conductor.bch>
    <cim:IdentifiedObject.name>Line A
       </cim:IdentifiedObject.name>
  </cim:ACLineSegment>
  <cim:Terminal rdf:ID="A_Te8">
    <cim:Terminal.ConductingEquipment rdf:resource="#A_A1" />
    <cim:Terminal.ConnectivityNode rdf:resource="#A_CN3" />
  </cim:Terminal>
```

```
  <cim:ConnectivityNode rdf:ID="A_CN3"></cim:ConnectivityNode>
  <cim:Terminal rdf:ID="A_T7">
    <cim:Terminal.ConductingEquipment rdf:resource="#A_B2" />
    <cim:Terminal.ConnectivityNode rdf:resource="#A_CN3" />
  </cim:Terminal>
  <cim:BusbarSection rdf:ID="A_B2">
    <cim:IdentifiedObject.name>Busbar 2
        </cim:IdentifiedObject.name>
  </cim:BusbarSection>
</rdf:RDF>
```

Appendix C
Exemplary Use Case According to an Extended IEC/PAS 62559 Template

The development and management of use cases is a challenge of great importance regarding the development of the Smart Grid, which incorporates several systems and stakeholders. Thus, it is subject of several projects and working groups. With use cases, the identification and management of requirements (see Chapter 2), system architecting (see Chapter 4), and related standardization can be supported.

Use cases basically identify actors (systems, components, persons, etc.) and their goals regarding functions of a particular system-of-interest, i.e. the Smart Grid or parts thereof. They outline scenarios which may occur when actors try to achieve certain goals with the considered system. A use case contains all relevant information required to achieve these goals and abstracts from specific technical solutions. Further explanations on use cases, their development and management can be found in Chapter 3.

This appendix shows an exemplary use case based on an example provided by the CEN/CENELEC/ETSI Smart Grid Coordination Group's (SG-CG) working group "Sustainable Processes" in a description of their template[1]. The template used in this chapter is based on the IEC/PAS 62559 template, work done at OFFIS, the SG-CG working group "Sustainable Processes", and IEC TC 8/WG AHG 4. The latter working group is currently concerned with the international, consensus-based standardization of a use case template for the energy sector on the basis of the IEC/-PAS 62559. With such a standardized template, it shall be ensured, that all relevant information is included and use cases become comparable and exchangeable. Also, people familiar with the template may find their way in development and application of use cases more easily.

[1] `ftp://ftp.cen.eu/CEN/Sectors/List/Energy/SmartGrids/Use%20Case%20Description.pdf`

1 Description of the Use Case

1.1 Name of Use Case

Use Case Identification		
ID	*Domain(s)*	*Name of Use Case*
UC-1012	Advanced Metering Infrastructure (AMI)	Read Remote Meter

1.2 Version Management

Version Management						
Changes/ Version	*Date*	*Name Author(s) or Committee*	*Domain Expert*	*Area of Expertise/ Domain/ Role*	*Title*	*Approval Status* *draft, for comments, for voting, final*
Version 0.2 Updated use case numbering & naming	2013-01-24	John Doe	Primary	AMI	Field Engineer	For Comments

1.3 Scope and Objectives of Use Case

Scope and Objectives of Use Case	
Related Business Case	Billing of energy consumption
Scope	Periodic collection of meter data through head end system. The detailed data transfer between meter and HES is excluded.
Objective	Remotely read Smart Meter data from Head End System

1.4 Narrative of Use Case

Narrative of Use Case
Short Description – *max 3 sentences*
This Use Case describes a part of the Advanced Metering Infrastructure (AMI). In particular it addresses the periodic collection of meter data through a Head End System (HES) when triggered from the Head End System Operator.
Complete Description
The Head End System (HES) Operator initiates a regular data collection process that is described in this Use Case. The time of the event for this regular data collection may differ, depending on contracts, legal billing requirements or needs from other processes that use the metering data. This Use Case consists of three main parts: 1. Activation of the meter reading scheme and data collection scheme at the Smart Meter level. The meter reading scheme generally includes the list of atomic billing data and metadata (as the time stamp of each billing period, the log of billing events, the schedule according to which these data are stored). The data collection scheme is the schedule according to which the stored data have to be pushed / pulled from the meter to the Head End System (HES). 2. Collection of metering data – which is not described in detail within this Use Case. This collection phase is repeated until the meter reading scheme is deactivated / changed (hence the meter reading scheme is not transferred every time billing data is required). 3. Deactivation of the meter reading scheme at Smart Meter level

1.5 General Remarks

General Remarks
This is an exemplary use case which is based on the description of the CEN/CENELEC/ETSI Smart Grid Coordination Group, working group "Sustainable Processes" template description example (see ftp://ftp.cen.eu/CEN/Sectors/List/Energy/SmartGrids/Use%20Case%20Description.pdf)

2 Diagrams of Use Case

Diagram of Use Case

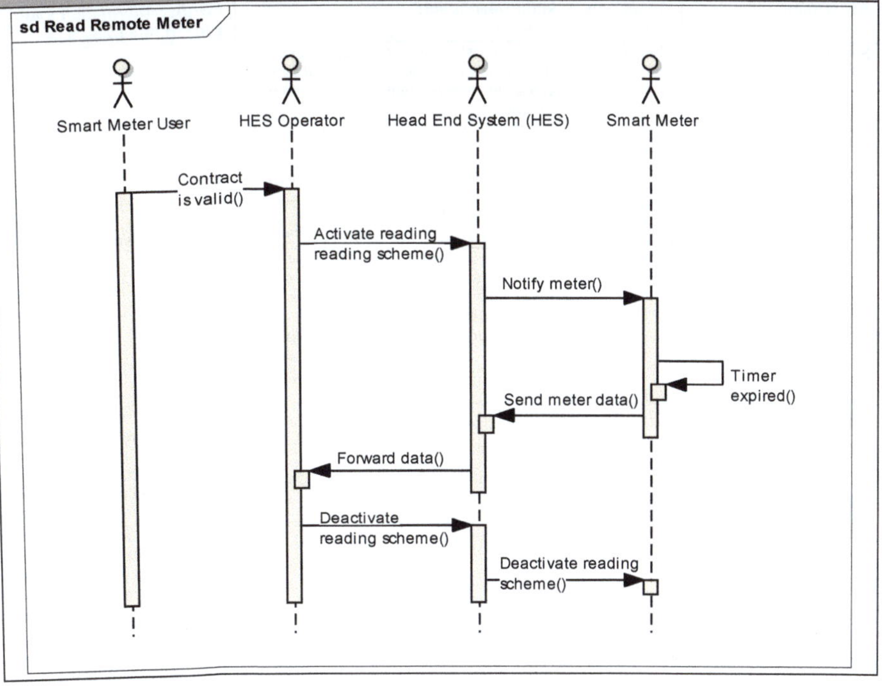

3 Technical Details

3.1 Actors: People, Systems, Applications, Databases, the Power System, and Other Stakeholders *(One table per "grouping")*

Actors			
Grouping (Community)		**Group Description**	
Service Provider		See SG-CG WG-RA, functional viewpoint	
Actor Name *see Actor List*	**Actor Type** *see Actor List*	**Actor Description** *see Actor List*	**Further information specific to this Use Case**
Head End System (HES)	System	Central data system collecting data via the AMI of various meters in its service area. The HES is part of the AMI and represents an interface for service providers.	The HES communicates via a WAN directly with the meters.
Head End System Operator	Person	Entity that offers services on a contractual basis, to collect metering data related to supply and to provide them to the relevant actor. The party is responsible for meter reading and quality assurance of the reading. It usually offers services on a contractual basis to provide, install, maintain, test, certify and decommission physical metering equipment related to a supply. In addition, the HES operation can offer services to aggregate metering data.	The periodic data collection occurs on a scheduled basis triggered by an external actor (e.g., a Timer). The communication can be based on either a pull or a push mechanism.

Actors			
Grouping (Community)		*Group Description*	
Customers		See SG-CG WG-RA, functional viewpoint	
Actor Name *see Actor List*	*Actor Type* *see Actor List*	*Actor Description* *see Actor List*	*Further information specific to this Use Case*
Smart Meter (SM)	System	Meter with at least functionalities to provide and transmit metering data as well as to receive control commands via communication networks.	-
Smart Meter User	Person	Customer, whose meter data is collected	-

3.2 Preconditions, Assumptions, Post condition, Events

Use Case Conditions			
Actor/System/Information/Contract	*Triggering Event*	*Pre-conditions*	*Assumption*
Head End System Operator	Head End System Operator receives a request for periodic metering data for billing purposes.	Communication with the meter can be established. The meter reading scheme and data collection schemes are available at HES level.	-
Head End System Operator	-	-	There is a valid contract between Consumer & Head End System Operator for collecting meter data
Smart Meter User	-	-	There is a valid contract between Consumer & Head End System Operator for collecting meter data
Smart Meter	-	-	An AMI Meter/Device is installed at the premise and operational.

3.3 References / Issues

References						
No.	*Reference Type*	*Reference*	*Status*	*Impact on Use Case*	*Originator/ Organisation*	*Link*
1	Standard	EN 62056-31:1999 Ed. 1.0, Electricity metering – Data exchange for meter reading, tariff and load control – Part 31: Use of local area networks on twisted pair with carrier signaling	IS	To be used for realization	CEN/CLC TC13	-
2	Standard	EN 62056-42:2002 Ed. 1.0, Electricity metering – Data exchange for meter reading, tariff and load control – Part 42: Physical layer services and procedures for connection-oriented asynchronous data exchange	IS	To be used for realization	CEN/CLC TC13	-
3	Standard	EN 62056-61: 2006 Ed. 2.0, Electricity metering - Data exchange for meter reading, tariff and load	IS	To be used for realization	CEN/CLC TC13	-

		control - Part 61: Object identification system (OBIS)				
4	Standard	EN 62056-62:2006 Ed. 2.0, Electricity metering - Data exchange for meter reading, tariff and load control - Part 62: Interface classes	IS	To be used for realization	CEN/CLC TC13	-
5	Standard	EN 13757-1:2002 Ed. 1.0, Communication systems for meters and remote reading of meters – Part 1: Data exchange	IS	To be used for realization	CEN/CLC TC 294	-
6	Standard	EN 61334-5-1:2001 Ed. 2.0, Distribution automation using distribution line carrier systems – Part 5-1: Lower layer profiles – The spread frequency shift keying (S-FSK) profile	IS	To be used for realization	IEC TC 57	-
7	Standard	EN 61334-4-32:1996 Ed. 1.0, Distribution automation using distribution line carrier systems – Part 4: Data communication protocols – Section 32: Data link layer – Logical link control (LLC)	IS	To be used for realization	IEC TC 57	-

3.4 Relation to other Use Cases and Contacts for modification

Classification Information
Relation to Other Use Cases
Belongs to "Billing Cluster"
Level of Depth
High level Use Case
Prioritization
• Obligatory, must be supported by metering standards • To be finished in 2013 • Final details might be different from country to country
Generic, Regional or National Relation
Generally applicable in Europe, however country-specific adaptions, e.g., an extra access layer, may be required.
Viewpoint
Technical
Further Keywords for Classification
Smart Metering, Meter Reading

4 Step by Step Analysis of Use Case

Scenario Conditions					
No.	**Scenario Name**	**Primary Actor**	**Triggering Event**	**Pre-Condition**	**Post-Condition**
1	Normal Sequence	Head End System Operator	Head End System Operator is informed that metering data from the Smart Meter is needed.	Communication with the meter can be established. The meter reading scheme and data collection schemes are available at HES level.	Head End System Operator Received all required periodic metering data.
2	Error Management	Head End System Operator	Deadline for reading certain metering data has passed and the Head End System Operator has not received all required data.	-	Head End System Operator Received certain required metering data.

4.1 Steps – Normal Sequence

Scenario								
Scenario Name :		Normal Sequence						
Step No.	*Event*	*Name of Process/Activity*	*Description of Process/Activity*	*Service Type*	*Information Producer*	*Information Receiver*	*Information Exchanged*	*Requirements R-ID*
1	Head End System Operator is informed that metering data from the Smart Meter is needed	Activate reading schemes	Head End System Operator activates meter reading scheme and data collection scheme at HES level.	REPORT	Head End System Operator	HES	Activation-Message for meter reading scheme	*Unique ID, e.g.* MR-1
2	A meter reading scheme is activated at HES level.	Notify meter	HES informs Smart Meter about activation of meter reading scheme. Optionally: if meter data is to be pushed by the meter, HES informs the meter about activation of data collection scheme.	REPORT	HES	Smart Meter	Activation -Message for meter reading scheme	MR-2
3	Timer has expired	Read metering data	When the timer expires, a meter read operation is triggered. If Pull communication, Timer triggers meter read at HES level. If Push Communication, Timer triggers meter read at Smart Meter level.	GET	Timer	Smart Meter or HES	Timer event notification	MR-3
4	Meter read was triggered	Send metering data	Meter sends required metering data to HES.	GET	Meter	HES	Metering data	MR-4
5	HES received metering data	Forward metering data	HES forwards metering data to	GET	HES	Head End System Operator	Metering data	MR-5

			Head End System Operator, whenever the timer is active.					
6	Head End System Operator has deactivated meter reading scheme at HES level.	Deactivate reading schemes	HES informs Smart Meter about deactivation of meter reading scheme and/or data collection	CLOSE	HES	Smart Meter	Deactivation-Message for meter reading scheme	MR-6

4.2 Steps – Alternative, Error Management, and/or Maintenance/Backup Scenario

Scenario								
Scenario Name :		Error Management						
Step No.	*Event*	*Name of Process/Activity*	*Description of Process/Activity*	*Service*	*Information Producer*	*Information Receiver*	*Information Exchanged*	*Requirements R-ID*
1	Deadline for reading certain metering data has passed and the Head End System Operator has not received all required data.	Request on-demand read	Head End System Operator requests a remote on-demand read of Smart Meter.	REPEAT	Head End System Operator	HES	Activation-Message for meter reading scheme	MR-7
2	HES received ad hoc request from Head End System Operator	Read metering data	Read metering data from Smart Meter on-demand	GET	Smart Meter	HES	Metering data	MR-8
3	HES received metering data from Smart Meter	Exit on-demand read	HES deactivates on-demand read mode	CLOSE	HES	Smart Meter	Deactivation-Message for meter reading scheme	MR-9

5 Information Exchanged

Information Exchanged		
Name of Information Exchanged	*Description of Information Exchanged*	*Requirements to Information Data R-ID*
Activation-Message for meter reading scheme	Message representing the start of a transaction using a specific reading scheme	MR-10
Timer event notification	Notification regarding a timer event	MR-11
Metering data	The data read from the Smart Meter	MR-12
Deactivation-Message for meter reading scheme	Message representing the end of a transaction using a specific reading scheme	MR-13

6 Common Terms and Definitions

Common Terms and Definitions	
Term	*Definition*
Meter Reading Scheme	Specification of the reading process (e.g. resolution of meter data / frequency of meter reading)
On-demand read	Ad hoc request for meter data

Index

ACSI 123
AMI 23, 24, 26–29
ANSI C12 185

CC PP 139
CDC 124
CDPSM 104, 159
CEN 9
CENELEC 9
CIM 99, 183, 187
  CIS 110
  Data Model 101
  GID 111
  IOP Test 159
  IRM 111
  Message Exchange 105
  Message Structure 105
  Profile 103
  Profiles 159
  Serialization 104
  Testing 157
  Topology 108
CIMbaT 112, 202
CIMDesk 159
CIMSpy 112, 159
CIMTool 112
CIMug 100, 159
COSEM 182
CPSM 104, 159

Dänekas, Christian 15
DKE 4
DLMS 182, 183
DLMS/COSEM 183

eBIX 219
ebXML 213
EFET 216
eMIX 223
EN 13757-3 184
Energy Reference Model Catalog 90
  Classification 93
  Components 91
  Functional Reference Model 91
  Usage 90
ENTSO-E 217
ESB 104
ETP 4
ETSI 9

FEG 32
FSS Group 11

GeLi Gas 225
GID 159
González, José M. 15
GPKE 225
GWAC 20, 23

IEC 60050 121
IEC 61131-3 206
IEC 61850-1 121
IEC 61850-10 127
IEC 61850-2 121
IEC 61850-3 121
IEC 61850-4 121
IEC 61850-5 122
IEC 61850-6 122
IEC 61850-7-1 122
IEC 61850-7-2 123
IEC 61850-7-3 124
IEC 61850-7-4 124
IEC 61850-7-410 125
IEC 61850-7-420 126

IEC 61850-8-1 126
IEC 61850-9-1 126
IEC 61850-9-2 127
IEC 61968 100
IEC 61970 100
IEC 62056 182
IEC 62325 101, 222
IEC 62351 132
IEC 62443 135
IEC 62541 193
IEC SIA 7, 27, 74
IEC SMB SG 3 4
IEC TC 57 7
IEEE 7
Information Models 81, 83, 85
Information Sources 85
Information Systems 87
Inhouse Automation System 180
ISA 99 135
ISO/IEC 27000 Series 138
ISO/IEC/IEEE 42010 62

JWG 8

KNX 185

M-Bus 184
M/441 10, 180, 181
M/490 3, 8
Market Communication 101, 211, 225
Maturity Model 33
Metering 159

NERC CIP 136
NIST 5

OPC
  Classic 192
OPC UA 11 191
  Communication Services 196
  Address Space 194
  Base VariableTypes 195
  Complex ObjectTypes 195
  DataTypes 195
  Mapping 195
  Modeling 201
  Nodes 194
  ObjectTypes 194
  Profiles 199
  ReferenceTypes 194
  Security 199
  Technology Mapping 198
  VariableTypes 195

RAWG 11

Reference Model Catalog 85
  Energy Reference Model Catalog 90
Reference Models 84
Requirements
  Analysis 17, 23, 80, 85
  Documentation 17–19
  Elicitation 17, 18, 20, 21, 23, 32
  Engineering 15, 39, 60, 74, 75
  Management 17, 18, 31, 70 74
  Validation 17, 18
Roadmaps 3, 32

SDO 9
Security Metrics 142
Security Patterns 143
SG-CG 3, 9
SGAM 16, 21–23, 36, 45, 60, 66, 76
  Business Layer 22, 24
  Communication Layer 22, 28
  Component Layer 22, 28
  Function Layer 22, 26
  Information Layer 22, 27
SGMM 33, 76
SM-CG 180
Smart Energy Profile 186
Smart Grid
  Definition 3, 4
  Motivation 4
Smart Meter 179
SML 184
System-of-Systems 16

Technology Field 32, 33
TOGAF 19, 65
  ADM 68
    Architecture Vision 71
    Business Architecture 72
    Information Systems Architecture 72
    Preliminary 71
    Requirements Management 74
    Technology Architecture 74

UCMR 54
UMM 214
Use Case 3, 39
  Methodology 42
  Process 47, 49
  Repository 44
  Template 44
  Tool-Support 54
  UCMR 54
Uslar, Mathias 3

ZigBee 186

MIX
Papier aus verantwortungsvollen Quellen
Paper from responsible sources
FSC® C105338
FSC
www.fsc.org

If you have any concerns about our products,
you can contact us on
ProductSafety@springernature.com

In case Publisher is established outside the EU,
the EU authorized representative is:
Springer Nature Customer Service Center GmbH
Europaplatz 3, 69115 Heidelberg, Germany

Printed by Libri Plureos GmbH
in Hamburg, Germany